Meteorites of Alberta

Anthony J. Whyte

Meteorites of Alberta

 The University of Alberta Press

Published by

The University of Alberta Press
Ring House 2
Edmonton, Alberta, Canada T6G 2E1

Copyright © 2009, Anthony J. Whyte.

Library and Archives Canada Cataloguing in Publication

Whyte, Anthony J., 1951-
The meteorites of Alberta / Anthony J. Whyte.

Includes bibliographical references and index.
ISBN 978-0-88864-475-6

1. Meteorites--Alberta--History. I. Title.

QB755.5.C3W58 2009   523.5'1097123   C2008-907140-9

The University of Alberta Press is committed to protecting our natural environment. As part
of our efforts, this book is printed on Enviro Paper: it contains 100% post-consumer recycled
fibres and is acid- and chlorine-free.

The University of Alberta Press gratefully acknowledges the support received for its publishing
program from The Canada Council for the Arts. The University of Alberta Press also gratefully
acknowledges the financial support of the Government of Canada through the Book Publishing
Industry Development Program (BPIDP) and from the Alberta Foundation for the Arts for its
publishing activities.

The Edmonton Centre of the Royal Astronomical Society of Canada has used the opportunity
provided by the Province of Alberta's Gaming Commission with Centre-run casinos to purchase
copies of *The Meteorites of Alberta* for gratis distribution to Canadian universities, colleges,
technical institutes, science museums, planetariums and astronomical groups around the
length and breadth of the country.

Contents

Meteorites, being rocks from space, are solid samples of places that we cannot easily go. Ejected by collisions between the debris in our solar system and the surfaces of asteroids and planets, they represent a whole variety of planetary bodies, each with its own 4.5-billion-year geologic history. Meteorites may be from planetary bodies that are remarkably similar to the Earth, like Mars, or they may be from those that are completely alien, for which we have no analogous terrestrial environment.

Some meteorites preserve a record of geologic processes that occurred on these worlds, such as heating, metamorphism, and melting, that transformed the original materials into something that resembles the interior of the Earth and the lavas found at its surface. Other meteorites, having escaped significant heating, preserve the original materials—objects present in the disk of dust and gas that surrounded the early Sun before the planets formed, such as chondrules, calcium-aluminum inclusions, and even mineral grains from nearby supernovae. The study of these meteorites allows us to peer back in time to the very beginnings of the solar system, revealing to us a part of the solar system's history that we cannot otherwise access and something of our own origins.

Meteorites of Alberta details the scientific results of the study of the meteorites of Alberta by researchers all over the world. Places like Bruderheim and Abee (whose locations I admit I might not have otherwise known) have been connected with the world through scientific research on the meteorites that were found there. I suspect that the residents of Bruderheim, for example, never expected the meteorite that fell in such spectacular fashion on 4 March 1960 to be the object of so much scientific scrutiny.

I am struck by the sheer volume of scientific literature associated with some meteorites of Alberta, and the paucity associated with others. The University of Alberta Meteorite Collection, having grown over the past four decades to the second largest in Canada, exists to preserve meteorites from Alberta and elsewhere and to enable research on meteorites in the collection. With the advance of analytical tools, scientific questions that were previously unanswerable can now be considered—this is the ultimate argument for the long-term preservation of meteorites. Perhaps the future study of one of the Alberta meteorites will provide new insight into the formation and geologic evolution of the solar system.

Whereas the scientific story of a meteorite is one thing, the human story behind each meteorite is quite another. Every meteorite has such a story, perhaps involving a fireball witnessed by hundreds of people or simply the chance discovery of a strange rock while tilling a field or walking a dog. *Meteorites of Alberta* captures the story of each meteorite, involving ordinary, hardworking people who have been fortunate enough that the strange rock they found came from space.

Some meteorites, like Iron Creek, are perhaps more significant in the history of Alberta, but each story captured in this book represents a piece of the province's history that, for the most part, has been largely overlooked. Tony spent the better part of the summer of 2004 poring over our Department records, including correspondence between Bob Folinsbee and others. He has done an excellent job of synthesizing and capturing the details of the history of the meteorites.

The account in Chapter 18 of visiting the Cretaceous–Tertiary boundary layer in Alberta and the work that the author's daughter did to actually separate and study minerals from the impact that killed off the dinosaurs is inspirational. If you feel moved to do likewise, to go hunt for meteorites, or better yet, to study meteorites in school after reading this book, then I wish you all the best.

Chris Herd
Curator, University of Alberta Meteorite Collection
November 2006

"There is nothing more powerful than an idea whose time has come."
—Victor Hugo

Excepting only the Tagish Lake (British Columbia) carbonaceous chondrite and the Springwater (Saskatchewan) pallasite stony-iron, the suite of 16 Alberta meteorites represents the crown jewels of Canadian meteoritic science.

Books about meteorites generally fall into two categories: either they offer a broad overview, using meteorite examples from all over the world chosen to best illustrate different aspects of meteoritic science, or they take the form of terse, often unillustrated, catalogues of meteorites. Such books deal almost exclusively with the meteorites themselves. The human history—who found the meteorites, when, and how—is often mentioned only in passing, if at all.

The Meteorites of Alberta is first and foremost a historical book. The progress of meteoritic science is traced through the study of Alberta meteorites, although this approach has its limitations because the bulk of the research was done decades ago. However, I hope that the comprehensive coverage of the scientific literature and the extensive bibliography will prove useful to meteorite researchers. I also hope that the historical and human-interest accounts in the book will appeal to researchers and non-specialist readers alike.

The genesis of *The Meteorites of Alberta* can be traced back to when I helped my daughter Tanya with two of her science fair projects dealing with meteorites, mainly by finding and photocopying relevant scientific papers for her to read. I discovered that there were a great many such papers on Alberta meteorites (some of which are among the most thoroughly analyzed meteorites in the world), as well as numerous non-technical accounts full of fascinating stories. Not long after I thought of writing a book on Alberta meteorites, this idea began to take on a life of its own. Starting with the very first people I contacted for information and advice, the growing consensus seemed to be that this was a book that *needed* to be written. So many people and organizations have contributed so much in so many ways to this book that it is difficult to adequately thank them all.

I would like first to thank my wife Annie and my daughters Tanya and Andrea for all their love and forbearance, without which I could not have sustained the effort needed.

I must thank many friends and acquaintances, many of whom are members of the Royal Astronomical Society of Canada (RASC): Doug Hube, who provided early direction and support; Mel Rankin, who was the first person to actually put something in my hands; and Martin Connors, Alistair Ling, Franklin Loehde, and Mark Zalcik, who provided photographs. Sherry Campbell permitted the use of material from *Stardust,* the newsletter of the Edmonton Centre of the RASC. Wayne Barkhouse and Jay Anderson, past Editor and current Editor, respectively, of the *Journal of the Royal Astronomical Society of Canada*, arranged for the extensive use of material from that respected publication.

Boyd Wettlaufer, who found the Belly River meteorite over 60 years ago, chose to tell the true story of its discovery for this book. Thank you, Boyd.

For its massive, wide-ranging and unstinting support, the Department of Earth and Atmospheric Sciences at the University of Alberta deserves special recognition. Patricia Cavell (former Collections Manager), another early supporter of the book, arranged for me to access and examine the meteorite collection. Chris Herd (Curator of the University of Alberta Meteorite Collection) read two full drafts of the manuscript, providing invaluable and much appreciated scientific and editorial advice; arranged access to the collected papers of R.E. Folinsbee and D.G.W. Smith; and provided several photomicrographs of Alberta meteorites. Andrew Locock (current Collections Manager) photographed the Alberta meteorites anew for the book. Martin Sharp (Chairman of the Department of Earth and Atmospheric Sciences) also lent his support to the book.

I also wish to acknowledge my two anonymous reviewers for their conscientious and perceptive analyses of my original manuscript and their several excellent suggestions.

The complex transformation of *The Meteorites of Alberta* from manuscript into published book was ably managed by my editors, Michael Luski and Peter Midgley, and their colleagues at University of Alberta Press: Linda Cameron, Alan Brownoff, Cathie Crooks, Mary Lou Roy and Yoko Sekiya. Thank you.

I would also like to thank Henry Bland, Frannie Blondheim, Robert Folinsbee, Ahmed El Goresy, Jon Greggs, Art Griffin, Ian Halliday, Laura Hayward, Alan Hildebrand, Hazel Humm, Russell Kempton, Ken and Marion Knudsen, Larry R. Lines, James G. MacGregor, Jr., Kurt Marti, Jessica Norris, Fred Rinas, Allen Ronaghan, Douglas Schmitt, Dorian Smith, Dennis Urquhart,

Joseph Veverka, Stan Walker, John T. Wasson, Graham C. Wilson, Alan Zalaski and many others.

Finally, I am certain that all the contributors enumerated above would join me in thanking the Edmonton Centre of the Royal Astronomical Society of Canada for their very generous financial support, without which this book could not have been published. The Edmonton Centre is dedicated to stimulating interest in astronomy and related sciences, and supports individuals and institutions engaged in the study and advancement of astronomy. Edmonton Centre members Orla Aaquist, Alicja Borowski, David Cleary, Franklin Loehde, and Bruce McCurdy were instrumental in obtaining the centre's approval of a grant to the University of Alberta Press.

Anthony J. Whyte
Edmonton, Alberta
October 2007

Acknowledgements

Acknowledgements for permission to reproduce images may be found in the *Image Credits* section.

Grateful acknowledgement is made to the following for permission to reproduce printed material: Allen Ronaghan, for excerpts from his article in the *Alberta Historical Review* (copyright reverted to author); American Association for the Advancement of Science for excerpts from *Science*; American Geophysical Union for excerpts from *Journal of Geophysical Research*; American Physical Society for excerpts from *Review of Modern Physics*; The *Calgary Herald*, for excerpts from an article "Manitou's Meteorite," by Mark Lowey, 6 March 1993, page C10; the *Canadian Mineralogist* for excerpts from various articles; Canadian Society of Exploration Geophysicists, for excerpts from *Canadian Journal of Exploration Geophysics*; the *Edmonton Journal* for excerpts from the articles, "Meteorite seen falling northeast of Edmonton," 10 June 1952, page 30 and "Place Meteorite in Museum," 30 June 1960, page 12; Elsevier, for excerpts from *Analytica Chimica Acta, Chemical Geology, Earth and Planetary Science Letters, Geochimica et Cosmochimica Acta, Icarus, Nuclear Instruments and Methods in Physics Research B*, and *Physics of the Earth and Planetary Interiors*; James G. MacGregor, Jr., for extensive excerpts from his father's book *Behold the Shining Mountains*; Lunar and Planetary Institute for excerpts from the *Proceedings of the Lunar and Planetary Science Conference* series; Macmillan Publishers Limited for excerpts from *Nature*; Mineralogical Society of America, for excerpts from *American Mineralogist*; the Meteoritical Society for excerpts from *Meteoritics, Meteoritics & Planetary Science*, and *Meteoritical Bulletin*; Royal Astronomical Society of Canada for excerpts from the *Journal of the Royal Astronomical Society of Canada*; Royal Ontario Museum for excerpts from an anonymous document; Royal Society of Canada for excerpts from *Transactions of the Royal Society of Canada*; Sherritt International Corporation for an excerpt from *Sherritt Nickelodeon*; Sky Publishing for an excerpt from *Sky and Telescope*; Society of Exploration Geophysicists for excerpts from *Geophysics*; Springer Science and Business Media for excerpts from *Space Science Reviews*; Terra Scientific Publishing Company for excerpts from *Geochemical Journal*;

Turnstone Geological Services Limited for excerpts from a report; and Verlag der Zeitschrift für Naturforschung for excerpts from *Zeitschrift für Naturforschung*.

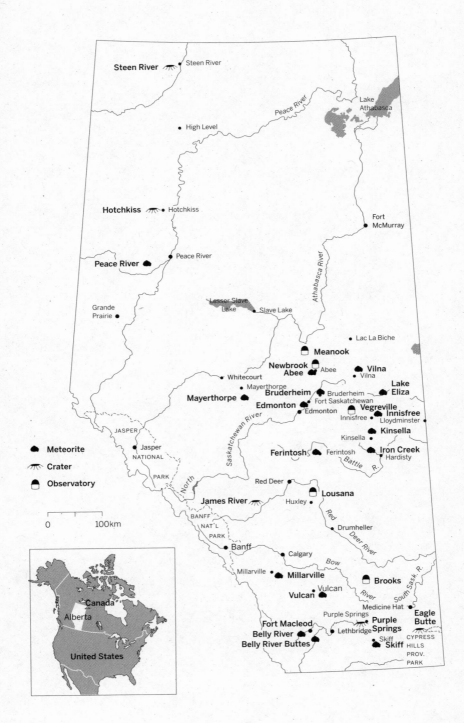

Steen River 〜 • Steen River

Peace River

Lake
Athabasca

• High Level

Hotchkiss 〜 • Hotchkiss

Fort
McMurray •

Peace River 🌩 • Peace River

• Grande
Prairie

Lesser Slave
Lake
• Slave Lake

Athabasca River

• Lac La Biche

Meanook 🏛 **Meanook**

Newbrook
Abee 🏛 • Abee

Vilna 🌩 **Vilna**
• Vilna

• Whitecourt

Lake
Eliza

• Mayerthorpe

Mayerthorpe 🌩 **Bruderheim** 🌩 Bruderheim
Fort Saskatchewan

Edmonton
• Edmonton

🏛 **Vegreville**
Innisfree

• Innisfree
• Lloydminster

JASPER
• Jasper

NATIONAL

Saskatchewan River

Kinsella • **Kinsella** 🌩

🌩 **Meteorite**
〜 **Crater**
🏛 **Observatory**

North

PARK

Ferintosh 🌩 Ferintosh

Iron Creek 🌩
Hardisty •

Battle R.

• Red Deer
🏛 **Lousana**

James River 〜
• Huxley

Red

Deer

BANFF

NAT'L

• Drumheller

PARK

0   100km

• Banff

• Calgary

Bow

River

Millarville • 🌩 **Millarville**

🏛 **Brooks**

Canada

Vulcan 🌩 • Vulcan

• Medicine Hat

Alberta

Purple Springs

Purple
🌩 **Springs**

Eagle
Butte

United States

Fort Macleod
Belly River 🌩
Belly River Buttes 🌩

• Lethbridge

〜

CYPRESS

• Skiff **Skiff** 🌩

HILLS

PROV.

PARK

South Sask. R.

Introduction

Interplanetary space is not empty. It is filled with rocky debris called *meteoroids*, ranging in size from specks of dust to mountain-sized chunks, travelling around the Sun in orbits.[1] Most of the smaller particles come from comets that hurtle around the Sun leaving a trail of dust. As the Earth orbits the Sun, it sweeps up an estimated 100 tonnes of this material every day. When a sand-grain-sized particle slams into the Earth's atmosphere at up to 40 km/s (about 144,000 km/h), it is heated to incandescence by friction with the air causing a bright streak of light in the sky called a *meteor*. Most meteoroids are completely burnt up by their fiery passage through the atmosphere, but a fist-sized or larger object will often survive to land on the ground (or in the water) as a *meteorite*.

The outer surface of a meteorite is often covered by shallow, oval pits or depressions. These are known technically as *regmaglypts* or *piezoglyphs* (Greek *piezein*, to press + *glyph*, to carve) and popularly as "thumb-prints." They are formed during the meteorite's flight through the atmosphere where the softer portions of its outer shell were "eroded" away, leaving small scooped-out places. The Iron Creek and Belly River meteorites illustrate this phenomenon particularly well.

A meteorite that is recovered shortly after its fall is witnessed, e.g., Abee and Bruderheim, is termed a *fall*. By far, the majority of meteorites are found, often by chance, long after they have fallen unseen; these are called *finds.*

Meteorites can be divided into three main groups: iron, sometimes referred to as *siderites* (from *sideros*, Greek for iron); stony-iron, also known as *siderolites*; and stony, or *aerolites*. The great majority (some 95%) of known meteorites are stony meteorites, consisting mostly of silicate minerals with lesser amounts of metal and sulphide; most of those are termed *chondrites* on the basis of texture (Table 0.1). These may be considered the sweepings of the solar nebula—material, which, for the most part, condensed in the dust cloud beyond the periphery of the nascent Sun. This material gathered together into a number of asteroidal bodies, subsequently broken apart in impact events, which much later furnished us with most of our supply of meteorites (a very rare few have come from the Moon and Mars). A minority of these bodies, being larger and/or hotter than the rest, were in large part melted, allowing the heavier fractions to sink and form a metallic core. These differentiated planetesimals appear to have had a silicate-rich mantle and perhaps crust, as is inferred for the Earth, floating above a core rich in nickel, iron, and other metals. The irons are thought to have originated from the cores of such bodies after they, too, were shattered by impact events in the early solar system, while the "slag" nearer the surface was the source of the so-called *achondrite* meteorites. Lastly, the thin interface between core and mantle may have supplied some of the rare stony-irons; there are no known Alberta stony-iron meteorites.

Although only about 4% of all known meteorites are irons, 5 of the 16 Alberta ones are irons (Table 0.2); they also make up a disproportionate number of finds, as opposed to falls. They are the easiest meteorites to recognize because they consist of almost solid metal (a nickel-iron alloy), which is very different from terrestrial rocks; on the other hand, stony meteorites, particularly those that have been weathered, are often indistinguishable from ordinary rocks. Iron meteorites are divided into three main groups based upon their structural properties that, in turn, reflect the proportion of nickel content. *Hexahedrites* consist mainly of *kamacite*, a nickel-iron mineral, and have a low nickel content of up to 6%.[2] *Octahedrites* consist of kamacite, *taenite* (another nickel-iron mineral) and *plessite* (an intergrowth of kamacite and taenite) in varying proportions; their nickel content ranges from 6% to 13%. The third group, the nickel-rich *ataxites*, contain as much as 50% nickel. Iron meteorites are further classified into groups I to IV by their chemical composition.

Stony meteorites are subdivided into two smaller groups according to the presence or absence of *chondrules,* which are small, spherical, partially crystallized droplets of minerals embedded in glass. In 1863, a German scientist, Gustav Rose, observed that chondrules were abundant in some stony meteorites, but not in others, and named them *chondrites* and *achondrites*, respectively.[3]

Chondrites are classified still further by their chemical and mineralogical composition. Mineralogists have identified a variety of familiar minerals common to both meteorites and the Earth. These include olivine, pyroxene, feldspar, chromite, and, rarely, quartz and microscopic diamonds. On the other hand, some meteoritic minerals as *schreibersite, daubreelite,* and *wadsleyite* do not occur naturally on the Earth. It should be stressed that, although unusual combinations of known elements are present in meteorites, no new elements have been discovered in meteorites.[4] The most common, or ordinary,

Table 0.1. Summary of recovered meteorites.

Category and class	Falls	Finds	Total	%
Stony	**940**	**20,574**	**21,514**	**95.5**
Ordinary chondrites	739	13,526	14,265	
Carbonaceous chondrites	36	525	561	
Enstatite chondrites	15	186	201	
Other chondrites	31	132	163	
Achondrites	78	532	610	
Ungrouped and unclassified[a]	41	5673	5714	
Stony Iron	**12**	**104**	**116**	**0.5**
Iron	**48**	**817**	**865**	**3.85**
Unknown	**5**	**7**	**12**	**0.05**
Total[b]	**1005 (4.5)**	**21,502 (95.5)**	**22,507**	

Note: Numbers of meteorites are up to the end of December 1999; as of 12 September 2006, the Meteoritical Society's database gives a total of 33,531 meteorites with valid names and 2207 with provisional names. The table is compiled from data on the Natural History Museum website (www.nhm.ac.uk/research-curation/projects/metcat/), taken from M.M. Grady (2000).

[a]Ungrouped meteorites have been examined and found not to belong to a specific group; unclassified meteorites have not been examined.

[b]Percentages of the two types of meteorites are given in parentheses.

chondrites are classified as *H,* meaning high total iron content; *L,* meaning low total iron content; and *LL,* meaning both very low total iron content and very low metallic iron. The rare *C,* or carbonaceous, chondrites (of which there are no Alberta representatives) are carbon-rich; while the equally rare and peculiar *E,* or enstatite, chondrites consist largely of the mineral *enstatite*, which is a variety of pyroxene with relatively little or no iron.[5]

Two of the most common minerals present in chondrites are olivine and pyroxene, both iron-magnesium silicates. Typical olivine is what is known as a *solid solution series*; that is, it may assume compositions ranging between two pure "end-members," one iron-rich (*fayalite*, Fa) and one magnesium-rich (*forsterite*, Fo). In ordinary chondrites, the olivine has a majority of the magnesium-rich end-member and its composition may be expressed, for example, as Fa_{20}, meaning that in this instance it contains 20% of the fayalite molecule, 80% of the forsterite molecule. Pyroxenes are more complex solid solutions, with three end-members (rich in iron, magnesium, and calcium) required to adequately describe most common varieties. Two varieties of pyroxene, *bronzite* (containing 10%–20% ferrosilite [$FeSiO_3$]) and *hypersthene* (containing 20%–30% ferrosilite), were once used to classify chondrites. In older reports and books on

Table 0.2. Summary of Alberta meteorites.

Meteorite	Type	Class	Fall or find	Date	Mass (kg)
Abee	Enstatite chondrite	EH4	Fall	1952	107
Belly River	Ordinary chondrite	H6	Find	1943	7.9
Belly River Buttes	Ordinary chondrite	L6	Find	1992	1.5
Bruderheim	Ordinary chondrite	L6	Fall	1960	303
Edmonton (Canada)	Iron (hexahedrite)	IIA	Find	1939	17.34
Ferintosh	Ordinary chondrite	L6	Find	1965	2.201
Innisfree	Ordinary chondrite	LL5	Fall	1977	4.58
Iron Creek	Iron (octahedrite)	IIIA	Find	1869	145.85
Kinsella	Iron (octahedrite)	IIIB	Find	1946	3.72
Lake Eliza	Ordinary chondrite	H(?)	Find	2005	0.34
Mayerthorpe	Iron (octahedrite)	IA	Find	1964	12.61
Millarville	Iron	IVA (anom.)[a]	Find	1977	15.636
Peace River	Ordinary chondrite	L6	Fall	1963	45.76
Skiff	Ordinary chondrite	H4	Find	1966	3.54
Vilna	Ordinary chondrite	L5	Fall	1967	0.00014
Vulcan	Ordinary chondrite	H6	Find	1962	19

Note: Names given in italics are provisional names and need to be published in the *Meteoritical Bulletin* to become official. The table was compiled from data on the Turnstone Geological Services Ltd. website (www.turnstone.ca/canamet3.pdf), taken from G.C. Wilson (2006).

[a]Anomalous.

meteorites, the terms *olivine-hypersthene* and *olivine-bronzite* are often encountered, but these appellations were discontinued as being too confusing and were replaced with "group L" and "group H," respectively, in modern chondrite classification schemes.[6]

As we shall see later in this book not just scientists are interested in finding meteorites; a great many meteorites are found by ordinary people who know what to look for or, at least, are able to recognize the unusual. Andrew Locock (University of Alberta) has developed a short list of questions to help individuals identify possible meteorites:

1. Does the specimen feel unusually heavy for its size? (Yes = possible meteorite). Many meteorites, particularly iron meteorites, are quite dense and feel heavier than most rocks found on Earth.
2. Does the specimen attract a magnet? (Yes = possible meteorite). Almost all meteorites contain some iron-nickel metal and attract a magnet easily.

3. Can you see gray metal specks shining on any broken surface of the specimen? (Yes = possible meteorite). Most meteorites contain at least some iron-
nickel metal. These flecks may be seen glinting on a chipped surface.

4. Does the specimen have a thin black crust on its outer surface? (Yes = possible meteorite). When a meteor falls through the Earth's atmosphere, a
very thin layer on the outer surface of the rock melts. This thin layer is called
the fusion crust. It is usually black and has the texture of an eggshell.

5. Does the specimen appear to have thumbprints or dents (regmaglypts) on
its surface? (Yes = possible meteorite).

6. Does the specimen have any holes or bubbles in it? (No = possible meteorite). Meteorites do not have holes or bubbles. Slag from industrial processes
usually has holes or bubbles.

If the answers to questions 1 and 2 are No, then the rock is almost certainly
not a meteorite. If the rock is actually a meteorite, then the answers to most of
questions 1 through 5 should be Yes and to question 6 should be No.[7]

The importance of determining the abundance of different elements in
meteorites has long been recognized, and more research has been done on this
aspect of meteoritics than any other. A great many elemental analyses, covering
virtually the whole periodic table (see Appendix) and involving thousands of
meteorites, have been published; much of this data has been gathered together
in review papers and monographs.[8] The analysis of meteorites, particularly
chondrites, has proven problematic because of the inhomogeneous distribution of silicate, metal, and sulphide *phases* in chondrules, matrix, veins, and
breccia fragments and the difficulties presented by chemical separations of
elements, not to mention problems of contamination. This causes considerable uncertainty in comparing analytical results for different samples, even
from the same meteorite; as a result most elemental abundances in chondrites
are reported as mean *bulk compositions*. However, meteoriticists' interest in the
micro-scale distribution of elements in meteorites has helped drive the development of ever higher resolution analytical tools.

Developed after World War II, neutron activation analysis (NAA) is a nondestructive, extremely sensitive, accurate, and rapid method for measuring elemental abundance in meteorite and other specimens. Samples are irradiated,
or *activated*, by thermal (slow) neutrons from a nuclear reactor, which causes
various elements present to become radioactive. The activated samples are
then placed in a sensitive gamma-ray scintillation spectrometer, which counts
and analyzes the characteristic flashes of gamma radiation emitted by the
radioisotopes of different elements as they decay. If chemical separations are

done to samples after irradiation to remove radioisotopes that may cause interference in measuring elements of interest, the technique is called *radiochemical* neutron activation analysis (RNAA). It is also possible to simultaneously measure more than 30 elements by purely instrumental procedures by what is commonly called *instrumental* neutron activation analysis (INAA). In most cases, gamma rays are measured *after* the end of the irradiation by neutrons; this is called *delayed gamma* neutron activation analysis (DGNAA). If, however, the radioisotopes of interest are very short-lived, gamma rays must be measured *during* the irradiation; this is called *prompt gamma* neutron activation analysis (PGNAA).

Developed during the 1960s, *electron microprobe analysis* (EMP) has the advantage of not requiring a nuclear reactor and has become the analytical tool of choice. It is performed by detecting and counting the fluorescent X-rays produced by bombarding a sample with high-voltage electrons from an instrument similar to an electron microscope.

A meteorite has several ages: how long ago it was formed, the so-called *formation* or *solidification age*; how long ago its parent body was disrupted by collision with another asteroid, the *gas retention age;* how long it has been exposed to cosmic radiation in space, the *exposure age*; how long it has lain on the ground before being found, *the terrestrial age.*

From its formation in a parent body to its fall to Earth, a meteorite is exposed to cosmic radiation in space. Cosmic rays are not electromagnetic radiation like X-rays, gamma rays, or visible light; they are actually highly energetic particles of matter (mostly helium nuclei, protons, and electrons), which travel across space at close to the speed of light. The lowest energy cosmic rays come from the Sun; higher energy ones originate in supernovae and pulsars within the galaxy; and those with the highest energy of all may be extragalactic in origin. When these so-called primary cosmic rays collide with the atoms in a meteoroid, for example, the resulting *spallation* reactions split off, or spall, parts of the atomic nuclei, transmuting them into different *spallogenic* (or *cosmogenic*) elements, which may be radioactive, or unstable, and decay into still other elements. The high-energy particles such as neutrons and protons (secondary cosmic rays) emitted in the course of this process then collide with other nuclei, causing a cascade of similar reactions. The amount of a cosmogenic radionuclide, e.g., ^{53}Mn or ^{26}Al, measured in a given meteorite can tell us much about the nature of the cosmic ray flux in space. It can also tell us much about the history of the meteorite from its formation in a parent body and subsequent exposure to cosmic radiation in space, to its fall to Earth.

The longer a meteorite is exposed to the Earth environment (its terrestrial age) before it is found, the more its structure and chemical composition are altered by weathering, corrosion, and contamination. A number of techniques have been used to determine the terrestrial age of meteorites including measuring radionuclides and, more recently, *thermoluminescence*. Thermoluminescence (TL) dating is based upon the fact that almost all minerals emit light when they are heated. Energy absorbed by a mineral from ionizing radiation frees electrons to move through the crystal lattice of the mineral and some are trapped at imperfections in the crystal. These trapped electrons are freed, emitting light as a result, when the mineral sample is heated under laboratory conditions. The light is measured with a sensitive photometer. The ionizing radiation can be in the form of cosmic rays that meteorites are naturally exposed to in space. As soon as a meteorite falls to the ground, it is largely protected from further cosmic ray bombardment by the Earth's magnetic field and atmosphere, and its natural TL begins to decrease or *decay*. Therefore, the terrestrial age of the meteorite can be determined from the level of the remaining natural TL. The TL method is both quicker and requires much smaller samples than methods that measure radionuclides. It can be used to rapidly screen large numbers of meteorites—thousands have been found in Antarctica—to select candidates for further study by other techniques such as NAA and EMP.[9]

I would like to explain to the reader how this book is organized. Each meteorite (except Lake Eliza, the most recent) gets its own chapter, each of which begins with a historical account of the meteorite ("History")—who found it, and when and how—full of human interest stories. Each chapter then continues with a summary of the scientific investigation of the meteorite ("Science"); in some of the larger chapters, this section is further subdivided into specific topics, e.g., "Fall Phenomena and Astronomical Interpretation" and "Mineralogy and Textures." These chapters are arranged by meteorite group: iron (Part I) and stony (Part II); within each part the meteorites are arranged alphabetically, except for Abee. Abee is listed last in Part II after the ordinary chondrites, because it is one of the rare enstatite chondrites. These meteorite chapters are followed by chapters on the all but forgotten story of Alberta's meteor observatories, the ongoing search for meteorites, and the death knell of the dinosaurs; plus, since meteorites continue to fall on Alberta and research goes on unabated, there is a final chapter covering recent developments.

With this brief introduction to meteoritic science—more information will be presented throughout the book—it is time to begin our story of the meteorites of Alberta.

One

Iron Meteorites

The Edmonton (Canada) Meteorite

History

On 9 March 1939, Professor John A. Allan, of the Department of Geology at the University of Alberta, addressed the monthly meeting of the Edmonton Centre of the Royal Astronomical Society of Canada. The topic of his lecture was "Meteorites." He began with a brief discussion of the origin and probable age of the Earth, then gave a detailed description of the appearance and characteristics of meteorites. Finally he exhibited the Edmonton meteorite (Fig. 1.1), which he was "then engaged in examining." This, then, was the public and scientific debut of the Edmonton meteorite. A report about the Edmonton meeting in the *Journal of the Royal Astronomical Society of Canada* created a brief flurry of interest.[1] Peter M. Millman, who edited the *Journal*'s regular "Meteor News" feature, wrote to Dr. Allan asking for an article about the meteorite, "preferably with several illustrations."[2] Allan said he would be very glad to prepare a short article for the *Journal* "just as soon as I have the information I require on the specimen."[3] Millman also reported the meteorite to the Society for Research on Meteorites, which maintained a catalogue of meteorites; it was number 241 on their list.[4] Dr. Frederick Leonard from the Department of Astronomy of the University of California, Los Angeles, also wrote, saying: "I had never heard before of the Edmonton meteorite and I should like to know more about it."[5] Allan told Leonard that he hoped "some time during the next few months to present a paper for publication."[6] Neither the article nor the paper ever materialized.

The Edmonton meteorite is an iron meteorite (Fig. 1.1); this much was known from the start, but further classification had to await detailed examination. Dr. Allan began his study of the Edmonton meteorite somewhat at a disadvantage, not knowing who had found it, or when, or where. Years later, in a report written after his retirement, Allan stated bluntly: "Details on the discovery of this meteorite are not known."[7] His best efforts had only been able to establish that the meteorite had been plowed up on a farm about 6–10 km north of the City of Edmonton and about 3–6 km east of the road to the Namao

Fig. 1.1. The Edmonton meteorite, showing its faceted and cubic or boxy shape.

Airport; this would be in the eastern half of Township 54, Range 24, west of the Fourth Meridian. This location corresponds to approximately latitude 53°40′N and longitude 113°25′W.

Allan's report continued:

> Apparently this specimen was plowed up in the early [1930s], and for some years was regarded as a curio, and finally was brought into the Industrial Laboratories at the University of Alberta. In 1938, it was given to me in the Department of Geology, and I had the specimen photographed and analyzed. Polished sections were examined in the Metallurgy Department.[8]

Science

Shortly after receiving the meteorite, Allan had it measured (mass, 7.34 kg; specific gravity, 7.56; dimensions, about 15 × 12 × 10 cm) and photographed.

Fig. 1.2. A polished and etched section of the Edmonton meteorite showing two distinct sets of diagonal Neumann lines. On the actual specimen, cubic cleavage planes can also be seen running parallel to the edges. Note also the thin heat-affected zone around the periphery. The dark area in the centre is the shadow from the camera.

J.A. Kelso, Director of the Industrial Laboratories at the University of Alberta, made a chemical analysis of a fragment sawn from the edge of the specimen and found that it consisted mainly of iron (93.75%), nickel (5.89%), phosphorus (0.21%), and sulphur (0.01%).[9] The meteorite was then sectioned, polished, and etched by R.M. Scott of the Department of Metallurgy.

After a cut section of an iron meteorite has been polished, the flat surface, except for possible inclusions, is mirror-like and resembles stainless steel. It appears to be remarkably uniform and uninteresting, but this appearance is misleading. When such a polished nickel-iron surface is etched with a weak solution of nitric acid, a characteristic and beautiful structural pattern develops. Like all hexahedrites, the Edmonton meteorite exhibits a pattern of thin straight lines called *Neumann lines* (Fig. 1.2).

Fig. 1.3. The 12 possible directions of Neumann lines oriented along the face of a cube.

The Neumann lines reflect the inner crystalline structure of the Edmonton meteorite, which is in fact a single, large, cubic crystal of kamacite. The directions of individual sets of lines are not random but pass along diagonals to the corners and midpoints of the edge of a cube. Thus, the sets of lines may assume up to 12 possible directions (Fig. 1.3). Two such sets of lines are clearly visible in Figure 1.2 and more are discernible on the actual specimen.[10]

Hexahedrites cleave, or split, relatively easily along cubic planes, which may account for the Edmonton meteorite's shape.[11] V.F. Buchwald in his *Handbook of Iron Meteorites* described it as being "partly bordered on four sides by cubic cleavage planes so that the exterior appearance approximates a truncated box."[12] He went on to say:

> the exterior cube form as we see it now probably closely resembles the shape that the mass had when circling in [the] cosmos, except for the truncated, rounded portions which were ablation-sculptured during penetration of the atmosphere. In other words, it appears that Edmonton at an early time was dislodged from its 'matrix' in the form of a boxlike, cubic cleavage fragment, which was only little modified during the atmospheric flight.[13]

Cleavage planes can be seen on the polished surface—they are not very well reproduced in Figure 1.2—running parallel to the exterior surfaces; they stand out because they are easily corroded and are now filled with oxide (rust).

Buchwald noted that the Edmonton meteorite is among the softest meteorites ever recorded; the microhardness of the kamacite is only 125 ± 5, i.e., about the same as calcite, increasing to 175 ± 20 in the heat-affected outer zone or rim.[14] The meteorite, he says, "must have been thoroughly annealed during its cosmic sojourn." Annealing is a process in which, by heating and then cooling a metal, the metal's crystalline microstructure is changed, and it is softened.

The meteorite consists of kamacite containing an abundance of small phosphide particles, with scattered 1–4 mm diameter troilite inclusions and needle-shaped schreibersite masses. Adjacent to the troilite inclusions and the schreibersite masses, there are no small phosphide particles. As a result of these varying concentrations of phosphide particles, the polished and etched surface of the meteorite has a mottled appearance.[15] Also visible around the periphery of the meteorite is a distinct 1.0–1.5 mm continuous heat-affected zone or layer caused by the fiery passage though the atmosphere. As we have seen, Allan never wrote an article or paper on the Edmonton meteorite and in 1953 Peter Millman wrote again, this time to Professor Robert (Bob) E. Folinsbee at the University of Alberta, urging:

> Since to my knowledge the Edmonton Meteorite has not been described in the scientific literature, I think it very important that some form of publication should appear...I am anxious that our Canadian meteorites should not be neglected.[16]

To encourage further research on Alberta meteorites and to build up the University of Alberta's meteorite collection, Folinsbee began exchanging meteorites with other museums. In 1956, he exchanged a 0.5 kg section of the Edmonton meteorite with the United States National Museum (USNM) in Washington, DC, for a slice of the Dimmitt, Texas, chondrite.[17] The USNM later provided a small piece from this section for a 1963 study of the helium (He) content of 34 different iron meteorites, including the Edmonton, by C.A. Bauer (Pennsylvania State University).[18]

Most of the helium found in iron meteorites was produced by the action of cosmic radiation before the meteoroid entered the Earth's atmosphere. Cosmic rays are high-speed particles—atomic nuclei or electrons—that travel throughout the Milky Way galaxy. Some of these particles originate from the Sun, but most come from sources outside the solar system and are known as galactic cosmic rays (GCRs). Most of the GCRs are protons (nuclei of hydrogen atoms), some are alpha particles (He nuclei), and a few are electrons and nuclei of heavier atoms. The interaction of GCRs with meteoritic material produces

Table 1.1. Major and minor elements in the Edmonton meteorite.

Method	Fe (mg)	Co (mg)	Ni (mg)	P (mg)	Cu (µg)	S (µg)	Cr (µg)	Ga (µg)	Mn (µg)	Ir (µg)
Wet chemical analyses	937.5	0.00	58.9	2.1	100					Trace
Atomic absorption spectrometry			53.7							
NAA								60.4		34
Ion microprobe mass spectrometry				0.37	160		120		40	
RNAA					100			70	15	37
EMP	941.3[a]	45[a]	52.6[a]	15.35[b]						

Note: Concentrations are in units per gram of meteorite. Method abbreviations are as follows: NAA, neutron activation analysis; RNAA, radiochemical neutron activation analysis; EMP, electron microprobe.

[a]In metal phase, inner zone.

[b]In phosphide.

[c]Determined photometrically by phenylfluorone.

a large variety of stable and unstable (radioactive) atoms including the helium isotopes, ^{3}He and ^{4}He. The total concentration of this so-called cosmogenic helium is proportional to the intensity of the cosmic radiation and the length of time during which the meteoroid was exposed. The ^{3}He/^{4}He isotope ratio of the cosmogenic helium depends on the size of the meteoroid and the position of the test sample within the meteoroid.

Bauer extracted the helium from the meteorite samples by heating them in a vacuum and analyzing the gas given off in a mass spectrometer. This method is so sensitive that only very small (0.10–0.30 g) samples of meteorite are needed; this sensitivity also means that the measurements need to be calibrated for helium (principally ^{4}He) leaking into the mass spectrometer from the atmosphere. From the total helium (^{3}He + ^{4}He) and the ^{3}He/^{4}He ratio, Bauer was able to calculate the relative rate of ^{4}He production and, from this, to estimate the exposure age or the length of time each meteoroid floated in interplanetary space before falling on the Earth. His estimated exposure ages for the various meteorites ranged from 40 Ma to over 1.5 Ga. The Edmonton meteorite's helium concentration (0.108 cm^3/g × 10^6 [at standard temperature and pressure]) was the lowest measured. Its ratio (0.117), also the lowest, implied that not all the helium was of cosmogenic origin, and therefore, the assumptions on which Bauer based his age estimations did not hold. Thus, he was not able to estimate an exposure age for the Edmonton meteorite.

The structural classification of iron meteorites into hexahedrites, octahedrites, and ataxites has been widely used since the late 1800s, but attempts to further subdivide irons by their chemical composition have not been as

Table 1.1. *(continued)*

Method	Ge (µg)	Zn (µg)	Si (µg)	Na (µg)	As (µg)	Os (µg)	Re (µg)	Reference
Wet chemical analyses								Kelso 1939
Atomic absorption spectrometry								Wasson 1969
Ion microprobe mass spectrometry	172							Wasson 1969
IMMS		≤20	6.2	≤10				Weinke et al. 1979
RNAA	179[c]	0.44			4.8	19.8	2.3	Pernicka et al. 1979
EMP								Pernicka et al. 1979

successful. In 1957, John Lovering (California Institute of Technology) and his colleagues proposed dividing irons into four groups (I, II, III, and IV) based in part upon the meteorites' gallium (Ga) and germanium (Ge) content.[19] Lovering's paper is also of interest to us because, although he lists an "Edmonton" meteorite among the 88 he studied, it is not *our* Edmonton meteorite. The one he lists was plowed up on a farm near Edmonton, Kentucky, in 1942! It, too, is an iron meteorite of similar size (10.2 kg), although of a different type—an octahedrite of group IIIC.[20] Lovering did not state which Edmonton meteorite he meant and, unless you remember that the first Edmonton meteorite—ours—is a hexahedrite, this could be confusing.

In April 1966, John T. Wasson (University of California, Los Angeles) wrote to Folinsbee at the University of Alberta about his "exhaustive study of the Ga-Ge-Ni [gallium, germanium, nickel] classification of iron meteorites, which will ultimately result in the analysis of 150–200 irons."[21] He asked for samples of the Edmonton and Mayerthorpe meteorites, saying "I need a minimum of about 8 g for my present studies, but would like to have about 20 g to allow future analyses to be made on the same samples." Folinsbee sent a 20 g piece of each meteorite to him. Wasson and his colleagues attempted to refine Lovering's four groups (I–IV) by studying a larger number of iron meteorites in greater detail. As new groups were discovered these were assigned letters within each of the original Ga-Ge groups (IA, IB, IIA, IIB, IIC, IIIA, IVB, etc.). In the third of a series of papers on this study, Wasson reported his analysis of the Edmonton meteorite (Table 1.1).[22] The level of iridium (Ir) was among the highest measured in the 61 iron meteorites tested in this part of the study.

Table 1.2. Trace elements in the Edmonton meteorite.

Method	Ru (ng)	Au (ng)	Sn (ng)	Sb (ng)	Hg (ng)	Sc (ng)	Mg (ng)
Ion microprobe mass spectrometry							≤60
RNAA	880	610	160	80	≤10	≤10	

Note: Concentrations are in units per gram of meteorite. RNAA, radiochemical neutron activation analysis.

Iridium, a platinum-like element, is a *siderophile* (iron-lover), meaning that it is closely associated with iron. The inner iron core of the Earth, for example, is much richer in Ir than the outer crust; terrestrial crustal rocks typically contain only about one *part per billion* (1 ppb) Ir. Therefore, iron meteorites, with their relatively high Ir content, are believed to represent material from the inner cores of asteroids. The Edmonton meteorite is a single crystal of iron alloy, and this gives us a clue as to the nature of its parent object. Large crystals have to form very slowly, and this suggests that the iron core of the asteroid cooled extremely slowly, perhaps at rates of only a few degrees Celsius per million years. This, in turn, implies that the iron, to have cooled so slowly, must have formed under a thick, insulating layer of overlying rock.

In 1975, Edward Scott and John Wasson proposed a classification scheme for iron meteorites that groups together genetically related irons, i.e., those that formed in the same locality in the solar system and experienced a similar chemical and physical history.[23] Their scheme incorporated a variety of parameters that had been used at different times to classify irons, including more obvious ones such as chemical, structural, or mineralogical properties and less obvious ones, which were determined indirectly, e.g., cosmic exposure ages and cooling rates. They first consolidated the 15 groups so far identified, retaining only 8 of these as separate entities and creating 4 composite groups (by combining IA and IB into IAB, IIA and IIB into IIAB, etc.). The new classification scheme thus consisted of groups IAB, 1C, IIAB, IIC, IID, IIE, IIIAB, IIICD, IIIE, IIIF, IVA, and IVB. A comparison of these 12 groups suggested to Scott and Wasson that there are two types with very different histories: (1) the major groups IIAB, IIIAB, and IVA (11%, 32%, and 8% of all irons, respectively), probably IIC, IID, and IVB, and possibly also IC, IIIE, and IIIF; and (2) the large group IAB (19% of all irons), IIICD, and probably IIE. According to Scott and Wasson, the first type of groups apparently formed in the molten cores of asteroids, each group forming in its own parent body. Groups of the second type seem to have diverse formational histories, but unlike the first type, they were not once part of molten cores.

Table 1.2. (continued)

Method	Al (ng)	K (ng)	Ca (ng)	Ti (ng)	V (ng)	Reference
Ion microprobe mass spectrometry	≤120	≤10	≤10	≤30	≤160	Weinke et al. 1979
RNAA						Pernicka et al. 1979

In January 1969, Hans Voshage (Max-Planck-Institut für Chemie, Mainz, Germany) wrote to Folinsbee asking for a sample of the Edmonton hexahedrite to test for its cosmic exposure age.[24] It was almost August before Folinsbee sent an 18 g slice, apologizing for the delay; the meteorite had been on a travelling display. Ten years later, Ernst Pernicka and his colleagues (Institute of Analytical Chemistry, University of Vienna) published the results of their trace element and electron microprobe analyses of the Edmonton meteorite (Table 1.2).[25] They reported that the heat-altered outer zone of the meteorite is significantly enriched in cobalt (0.64%) and nickel (5.63%) compared to the inner zone (0.45% and 5.26%, respectively). This corresponds to a greater microhardness, as already observed by Buchwald.[26]

H.H. Weinke, R. Gijbels, R. Saelens, and W. Kiesl (University of Antwerp and University of Vienna) investigated the Edmonton meteorite and four other irons by ion microprobe mass analysis.[27] They measured the trace element content for several alkali metals, transition metals, and other elements (Table 1.2). They were especially interested in the silicon because of its possible role in the origin of hexahedrites.

It had been proposed by other researchers that enstatite chondrites (e.g., Abee; see Chapter 15) could be the starting material for hexahedrites. However, Weinke and his colleagues argued, the small Si content between 6 and 40 parts per million (ppm) of the hexahedrites analyzed in their study shows this mechanism to be impossible, because the high Si concentration of enstatite chondrites (up to 3.3% by weight) can be removed from the nickel-iron phase only under oxidizing conditions, which would shift the Fe/Ni ratio to lower values. Hexahedrites with their Ni content between 5.3% and 5.7% by weight could not be produced.

Summary

Edmonton is one of the most extensively analyzed Alberta iron meteorites. It was included in two major studies, by John T. Wasson's group, on the classification of iron meteorites by their chemical properties.

The Iron Creek Meteorite

History

James G. MacGregor, Alberta historian and gifted writer, described in his book *Behold the Shining Mountains* what it must have been like when, uncounted generations ago, the Iron Creek bolide flashed across the Alberta sky:

> Far out on the Neutral Hills, away south in the Hand Hills, or in the green forests of Beaver Hills, our Indians saw this great flash of light. It swept over Alberta, seeming so close to them that its light nearly blinded them and its roar nearly split their ear-drums. In the past, in dreams and visions, many individual Indians had heard the Great Spirit speaking in the sky or had seen him descending from the heavens. This time, however, whole camps at a time saw this manifestation of the Great One. Everyone wondered what might be its meaning. Humbly they discussed it. Did it portend good or evil? Had they sinned, or did the great light which flashed across the sky and came to earth far out on the prairie augur well? The wise Medicine Men, sly, but curious, communed with their familiar spirits. "We must go to the place where it fell," they announced.
>
> For days, from the forests and the parklands, from the foothills and the prairies, bands of reverent Indians closed in upon the spot on the banks of Iron Creek. What tribes were they? Who knows? For no one knows whether it fell three hundred or three thousand years ago. Whoever these Indians were, peace reigned amongst them or was imposed upon them by this dread manifestation of the Manito. Soon news reached those still approaching that it had been found on a hilltop. Hundreds, maybe thousands, flocked in and camped near this hill, while the wise men pondered.
>
> Large crowds were dangerous—such a trivial incident could set off a war. But in the prevailing awe, no man's hand was lifted against his neighbor. Large crowds were impractical, too, for how could the hunters forage for such a multitude? But the Great Spirit who had sent this miracle from the heavens—the Great Spirit who had led this crowd, also fed them. From the west, from the direction of Buffalo Lake, he sent buffalo by the

thousands grazing across the prairies. Their numbers were so great that the land looked black and appeared to move and the air vibrated with their bellowing. Here were enough buffalo to feed the world. Here was another miracle.

With fasting, grim visage and solemn chant, the wise ones carried on their councils. At last they perceived the wishes of the Great Spirit. "You are to dig in that hole in the ground," they instructed. "Dig through the charred grass and the burnt earth. Dig there till the message is revealed."

Deeply they dug, sweating in the sun, throwing up sand fused into glass and clay baked red. Hourly the hole deepened and at last they found it—this strange object sent by the Manito. Never had the most wrinkled patriarch seen its like before—black, shiny, pitted, and heavy beyond the weight of two other boulders. They lifted it, and set it on the hilltop. Tenderly the Medicine Men rubbed off the clay and the sand sticking to it; and lo, as the clay fell off, the silent multitude beheld the profile of the Great Spirit. There was the brow and the eye, the flat nose and the mouth and the chin. Surely this stern, inscrutable, uncompromising visage was an image of the Great Spirit. New awe descended on the crowd.

So they set him on the hill facing the south, the abode of the sun, turning his back on the north wind. Each warrior made him an offering, humbly asking that his prowess might be great and his days long in this lovely land. Each mother made him a prayer that her sons might grow to be like him, strong and stern of visage and great in battle; that her daughters might be fair as the morning breeze; that even in the biting blizzard he might send them food; and that always in this sun-kissed land he would keep them well.

For a thousand years the Manito Face looked south across the valley. For a thousand years wandering warriors him offerings of sweet grass, and solemn squaws left him morsels of buffalo, and all was well in the land. Glorious springs, glowing summers and gorgeous falls passed over the face of the parklands. Each spring the buffalo increased, till they covered the land. Each year the fresh breezes blew health to sun-tanned bodies. The Great Spirit had set his face towards his people. All was well in the land.

Then came the white man....[1]

On the morning of 11 September 1754, the explorer Anthony Henday became the first white man to enter Alberta and on September 24 he crossed Iron Creek near its confluence with the Battle River, about 10 km northeast of Hardisty. No mention, however, is made of the Manitou Stone in the

Fig. 2.1. The Iron Creek meteorite, or Manitou Stone.

published journal of his journey. However, the original copy of his journal was lost, so we will never know if the native people told him about the meteorite located nearby.

The earliest recorded reference to the meteorite was made by Alexander Henry, then at Paint Creek Post, about 20 km north of Marwayne. On Sunday, 2 September 1810, he wrote: "This afternoon four Crees with their families arrived from the Sarcee camp on the S. side of Battle River, at the Iron Stone."[2] Then in 1860, a Wesleyan missionary, Thomas Woolsey, writing to the Earl of Southesk, stated:

When with the Crees last August, I visited the locality renowned for having a large piece of iron there. In fact, an adjoining lake and a rivulet bear the respective designations of Iron Lake and Iron Rivulet. Well, there the iron is, as pure as possible, and as sonorous as an anvil, and weighs, I should judge, 200 lbs [91 kg]. It is on the summit of a mound, but whether it is a meteoritic phenomenon or indicative of iron in that section, I cannot say....[3]

After the explorers, the fur traders and the missionaries, came the tourists. In 1862, two Englishmen, the hale Dr. Walter Cheadle and his sickly, often querulous companion, Lord Milton, set off on the first ever trans-Canada journey undertaken simply "for pleasure." What is of interest to us is Cheadle's travel journal entry for Sunday, 12 April 1863: "Baptiste [Supernat, Cheadle's Métis guide]...tells us [of a] ; 'piece of iron found by Indian near Edmonton & placed many years ago on top of a hill, size of fist when placed there, now so large no one can raise it! place where originally found stream of red water.'"[4]

Cheadle's journal was not intended for publication (although it eventually was—in 1931) and Milton and Cheadle instead published an edited, more polished account of their epic adventure titled *The North-West Passage by Land*. In their published account Milton and Cheadle provided more details about the story:

The only thing which makes this tale worth mentioning, is that it obtains universal credence amongst the half-breeds [*sic*]. Many of them profess to have seen it, and one man told us he had visited it twice. On the first occasion he had lifted it with ease; on the second, several years afterwards, he was utterly unable to move it! The man most solemnly assured us that this was perfectly true.[5]

Milton and Cheadle also mention the meteorite as being "found many years ago, but within the memory of people still living,..."[6]

The Indians called the meteorite Manitou or Manito and were said to have left food and other trinkets before setting out on hunting or warlike expeditions. The missionaries in the area, fearing that it could prove difficult to convert the native people to Christianity if their "Manitou-stone" and its superstitions remained where it was, had the iron confiscated.[7]

The Methodist missionary George McDougall stated in a 1869 letter to Dr. Wood, of the Wesleyan Missionary Society in Toronto:

This beautiful stream [Iron Creek] derives its name from a strange formation, said to be pure iron. The piece weighs 300 lbs [136 kg]. It is so soft you can cut it with a knife. It rings like steel when struck with a piece of iron....For ages the tribes of Blackfeet and Crees have gathered their clans to pay homage to this wonderful manitoo. Three years ago, one of our people put the idol in his cart and brought it to Victoria. This aroused the ire of the conjurors. They declared that sickness, war, and decrease of buffalo would follow the sacrilege. Thanks to a kind Providence, these soothsayers have been confounded, for last summer thousands of wild cattle grazed upon the sacred plain.[8]

The "Victoria" referred to is not the one in British Columbia but, rather, the Methodist Mission established at Fort Victoria or Victoria settlement (now Pakan) on the North Saskatchewan River about 100 km northeast of Edmonton.

It is not entirely clear who was responsible for this act of pillage; George McDougall said "one of our people put the idol in his cart." A.P. Coleman, the first scientist to examine the meteorite, recorded that "Mr. David McDougall, at the instance of his father, the Rev. George McDougall, brought [it] in, by Red River cart...."[9] James G. MacGregor, in his *Behold the Shining Mountains,* said:

In 1869, John McDougall, on one of his many journeys to and fro across this great land, was shown this stone by the Indians. He recognized it as a meteorite and, incredible as the task may seem, had it hauled 150 miles [240 km] over the winding trail to the mission.[10]

The Alberta historian, Allen Ronaghan, found an unsigned document in the files of the Royal Ontario Museum that states:

The "Iron Creek" meteorite was brought to Victoria (now Pakan) about the year 1870 by one John Whitford (O-mah-chees), an English half-breed [*sic*]; from him it was purchased by Mr. David McDougall who presented it to his father, the late Rev. George McDougall,...[11]

Let us give the final word on this episode to Peter Millman and D.W.R. McKinley: "Maybe the first Canadian meteorite was not actually stolen in the strict sense of the word, but it certainly was taken from the Indians without their permission,..."[12]

It has become almost *de rigueur* to quote W.F. Butler's account of what happened next. In 1870, Captain (later Lieutenant General Sir) William Francis Butler received orders from the Lieutenant-Governor of Manitoba, Sir Adams George Archibald, to proceed on a mission to the Saskatchewan [River]. While returning from the West he passed, on 25 December 1871, through the village of Victoria and spent the evening of Christmas Day in Reverend George McDougall's house. He later recounted in his book *The Great Lone Land* that:

In the farmyard of the mission-house there lay a curious block of metal of immense weight; it was rugged, deeply indented, and polished on the outer edges of the indentations by the wear and friction of many years. Its history was a curious one. Longer than any man could say, it had lain on the summit of a hill far-out in the southern prairies. It had been a medicine-stone of surpassing virtue among the Indians over a vast territory. No tribe or portion of a tribe would pass in the vicinity without paying a visit to this great medicine: it was said to be increasing yearly in weight. Old men remembered having heard old men say that they had once lifted it easily from the ground. Now, no single man could carry it. And it was no wonder that this metallic stone should be a Manito-stone and an object of intense veneration to the Indian; it had come down from heaven; it did not belong to the earth, but had descended out of the sky; it was, in fact, an aerolite. Not very long before my visit this curious stone had been removed from the hill upon which it had so long rested and brought to the Mission of Victoria by some person from that place. When the Indians found that it had been taken away, they were loud in the expression of their regret. The old medicine-men declared that its removal would lead to great misfortunes, and that war, disease, and dearth of buffalo would afflict the tribes of the Saskatchewan. This was not a prophecy made after the occurrence of the plague of smallpox, for in a magazine published by the Wesleyan Society in Canada there appears a letter from the missionary, setting forth the predictions of the medicine-men a year prior to my visit. The letter concludes with an expression of thanks that their evil prognostications had not been attended with success. But a few months later brought all the three evils upon the Indians; and never, probably, since the first trader had reached the country had so many afflictions of war, famine, and plague fallen upon the Crees and the Blackfeet as during the year which succeeded the useless removal of their Manito-stone from the lone hill-top upon which the skies had cast it.[13]

How long the meteorite remained at the Victoria mission is also not clear. Millman stated that "it was brought east from what is now Alberta by Captain Butler in 1871;"[14] if he did, Butler does not mention it in his account. MacGregor said:

> Some years later [after Butler had seen it at Victoria] McDougall sent it to Winnipeg. The belief is that it was hauled there by Red River cart. It is possible, however, that if he kept it at Victoria until 1875, when the first steamer plied that far up the river, he could have sent it by water.[15]

The meticulous Ronaghan says only that, "By 1886, the meteorite was at Victoria College, Cobourg, Ontario, where it was studied and reported on by Dr. A.P. Coleman."[16] From there, in 1892, it was moved with the College to Toronto. It was later transferred to the Royal Ontario Museum, Toronto.

When the meteorite was first displayed at the Royal Ontario Museum, it was identified as coming from Iron Creek, Saskatchewan.[17] Although Iron Creek is today in Alberta, prior to 1905 when Alberta and Saskatchewan became provinces and their common boundary was shifted eastward (to 110°W), the location was indeed in the "District of Saskatchewan."[18] However, Iron Creek is more than 50 km long and pinpointing the original location of the meteorite is difficult. The aforementioned unsigned document found by Ronaghan stated that the meteorite was found "near the headwaters of the east branch" of [Iron Creek] and gives a geographical location of 53°0'N and 111°0'W.[19] Millman gave the latitude as 52°50'N and the longitude as 111°30'W.[20] M.H. Hey gave 53°N and 112°W in his *Catalogue of Meteorites*.[21]

In Ronaghan's opinion, the Iron Creek hill mentioned in several accounts lies " just a few miles north-east of Lougheed, and within a mile of Iron Creek [Fig. 2.2, below]. There are two hills, both on the same section of land, which could be Iron Creek hill. I tend to favor the more westerly one, known locally as 'Strawstack Hill.'" He continued:

> The hill is not high, but it is high enough to be seen for many miles along the length of Iron Creek. It can also be seen from high points like Ribstone Hill, southeast of Viking, and from Flagstaff Hill, south-west of Hardisty.
>
> This hill is just a little more than ten miles [16 km] from the Battle River "as the crow flies," close enough to be used as a landmark of the kind mentioned in Henry's journal. The Sarcee camp was likely in the area south of Hardisty, an area that was a favourite camping and hunting ground of the Sarcees, Crees and Blackfoot Indians. Water, firewood and game were all plentiful in the region....[22]

Amateur historian Bill Peters, Sr., of Calgary learned of the Iron Creek meteorite's existence in the mid-1960s, in a library book brought home by his son. Peters thought that the meteorite probably sat on a mound, overlooking two small lakes about 4 km northwest of Hardisty.[23] This location matches Thomas Woolsey's account (above). The lake and rivulet still exist today (Fig. 2.3), but they have lost their names.

In early 1970, Peters gave a talk on the Iron Creek meteorite to the Calgary Centre of the Royal Astronomical Society of Canada. His lecture sparked the

Fig. 2.2. This hill, looking north on Range Road 105 northeast of Lougheed, once may have been the home of the Iron Creek meteorite.

interest of amateur astronomer John Howell, and the two of them began a collaborative search for the site where the meteorite originally lay. Their search of historical documents and exploration of the countryside around Iron Creek, in many ways, paralleled those of Allen Ronaghan. Peters and Howell were led to an interesting document file at the Provincial Museum of Alberta in Edmonton by David Spalding, then the Head Curator of Natural History. The papers included a record of a conversation between Professor R.E. Folinsbee at the University of Alberta and Bill McDonald, a geologist, about a possible second, larger, and still buried Iron Creek meteorite. Howell became convinced that there was, indeed, a much bigger piece still buried, possibly near where Iron Creek joins the Battle River (Fig. 2.4).[24] Peters disagreed, saying: "There's no evidence of anything else. None whatsoever."[25]

In January 1964, McDonald, who was an experienced geologist, told Folinsbee that:

> the larger fragment of Iron Creek was still in place on the Battle River in 1900, and that he himself saw it as a child of about 7, and that his brother, who was 17, often saw it.
>
> It was located...on a little ridge which overlooks the Battle River. All around the Manito stone were bags of beads...left as oblations....

Fig. 2.3. This may be the "Iron Lake" and "Iron Rivulet" (in foreground, overgrown with vegetation) with a mound nearby (left centre) referred to by the missionary Thomas Woolsey in 1860 as the site of the Iron Creek meteorite.

About 1903, the stone disappeared. It was said that the Indians buried it, or carried it away into a lake. Bill says his brother estimated the weight at 1500 lbs. [680 kg], and does not believe the native people could have carried it far. It is said that they carried it to a lake—the closest is about a mile away, and tossed it in. Bill and his brother think they dug a hole and toppled it in and left it.[26]

Folinsbee thought it was "a very good bet for investigation." However, it was not until September 1971 that he was able to organize a small group to follow up McDonald's story. They were not able to find any beads to pinpoint the exact location. Using metal detectors they searched the ridge overlooking the Battle River but found only buried tin cans and pieces of an old iron stove.[27] At least two more recent searches have similarly failed to find a second Iron Creek meteorite.[28]

It is interesting to speculate, though, that the persistant and consistant stories told to Butler, and to Milton and Cheadle (see above: "...one man told us he had visited it twice. On the first occasion he had lifted it with ease; on the second, several years afterwards, he was utterly unable to move it!") might be explained by there being at least *two* meteorites. It is conceivable that the smaller, less deeply-buried one—the one we know about—was found first and was worshipped by the natives, and a more massive, more deeply buried (harder to excavate) one was

Fig. 2.4. Iron Creek, near its confluence with the Battle River, about 6.5 km northeast of Hardisty.

found later. A man might visit and heft the original meteorite, only to return years later to find the heavier one in place. It is even likely that not all native people would be aware of the newly found large meteorite but would continue to venerate the original Manitou Stone; they would, unwittingly, prompt the McDougalls to remove the latter. Other native people could have hidden the larger meteorite to prevent it from being stolen too.

Science

A.P. Coleman (Victoria College, University of Toronto, Cobourg, Ontario) was the first scientist to describe the meteorite (he called it "a meteorite from the Northwest"):

> In outline, this meteorite is irregularly triangular and much broader than it is thick. Its surface shows the usual rounded and pitted appearance. It consists of solid metal, with scarcely a trace of stony matter, and only a slight oxidation of the surface. The specific gravity of the metal is 7.784. An analysis gives the following results: Iron (91.33 per cent; nickel, 8.83 per cent; cobalt, 0.49 per cent; total, 100.66).

He gave its weight as "about 386 lbs" (175 kg).[29]

In 1907, O.C. Farrington reported that a cast of the Iron Creek meteorite had been received by the Field Columbian Museum in Chicago "through the kindness of the Geological Survey of Canada." He provided a very detailed description of the meteorite's shape and surface appearance, as seen from its cast (Fig. 2.5). The form, he said, is

> that of a low cone, 8-1/2 inches (22 cm) high and 22 inches (56 cm) in diameter. The outline of the base of the cone is an incomplete circle, an approximately straight contour cutting off one side so that only about three-fourths of the circle is present. The width of the mass in this direction is 17 inches (43 cm).... At one point where the straight side joins the circular outline there was, evidently, in the original mass, a prolongation perhaps of a few inches in length, which having formed the most convenient part of the meteorite for removal has been sawed off for purposes, doubtless, of analysis and distribution.[30]

Farrington adopted the name Iron Creek for the meteorite

> instead of the more usual one of Victoria on account of information received from Mr. Johnston of the Geological Survey that the small mission station of Victoria, from which the meteorite received that name, is one hundred and fifty miles [240 km] from the locality where the meteorite was found, and it is no longer known by that name, its present name being Papan [sic].[31]

We have seen how, when a polished section of an iron meteorite with relatively low nickel content (i.e., an hexahedrite), such as the Edmonton (Chapter 1), is etched in dilute nitric acid, a pattern of lines called the Neumann lines appears on the surface. If the same procedure is done to an iron meteorite of higher nickel content, a distinctive network of crisscrossing stripes, or *bands*, flanged by shiny narrow *lamellae*, appears on the surface (Fig. 2.6). This pattern is called the *Widmanstätten* structure, in honour of the Austrian scientist Count Alois von Widmanstätten, who described them in 1808.[32] The Widmanstätten pattern is characteristic of the most common type of iron meteorites, including Iron Creek, known as *octahedrites* on account of their octahedral (eight-sided) crystalline structure. If you look carefully at the Iron Creek meteorite (presently on display in the Royal Alberta Museum in Edmonton) you can see the Widmanstätten pattern through the thin reddish-brown crust.

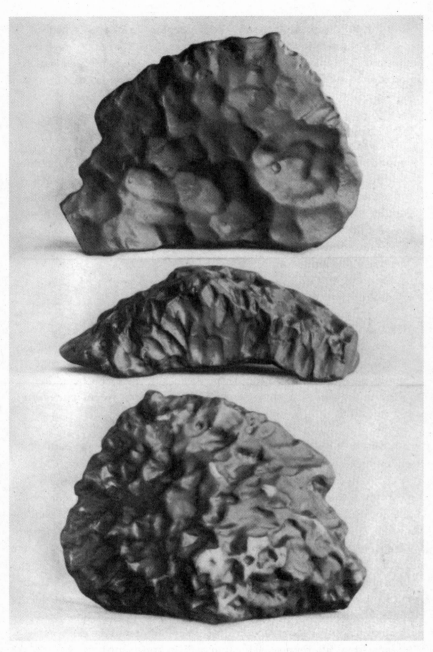

Fig. 2.5. A cast of the Iron Creek meteorite photographed in three views: rear (top); side (middle); and front (lower). Note where a sample has been sawed off (lower left of top view; lower right of lower view).

Fig. 2.6. A polished and etched section of the Iron Creek meteorite showing the Widmanstätten pattern.

0 15 mm

The Widmanstätten pattern is made visible by the differential etching of the meteorite surface by the acid solution. The wide bands are composed of kamacite (iron with only about 6% nickel) and etch easily, becoming dull (rough) in appearance. The thin lamellae flanking the bands consist of taenite (iron with 26%–30% nickel) and are not affected by the acid, remaining just as shiny after etching as before. The angles at which the kamacite bands crisscross depend only on the direction in which the meteorite is sectioned (cut). The width of these same bands, on the other hand, varies with the nickel content of the meteorite. Generally, the higher the nickel content, the thinner the bands are and the finer the Widmanstätten pattern; this offers a convenient method of classifying the octahedrites (Table 2.1). With a band width of 1.05 mm, the Iron Creek meteorite is classed as a medium octahedrite, Om.

The beautiful and distinctive Widmanstätten pattern is unique to meteorites and is never found in manufactured iron-nickel alloys of even identical chemical composition. It is generally accepted that iron meteorites represent material from asteroids large enough to have developed iron cores like the Earth's. As the molten iron core of such an asteroid, containing 6%–20% nickel (Ni), slowly cools down to about 1400 °C, the nickel-rich mineral taenite crystallizes out. At still lower temperatures (below ~900 °C) the low-nickel mineral kamacite begins to form, producing the Widmanstätten structure; for this crystallization to occur, the metal must cool extremely slowly. The estimated cooling rate, in the critical range between 700 and 500° C, is 1–10 °C per million (10^6) years.[33] It is for this reason that the Widmanstätten structure cannot be duplicated.

Lovering and his colleagues divided iron meteorites into four groups (I–IV) based in part upon the meteorites' gallium (Ga) and germanium (Ge) content.[34] This chemical classification was subsequently modified by other researchers

Table 2.1. Structural classes of iron meteorites.

Structural class	Symbol	Width of kamacite bands (mm)	Nickel content (% by mass)	Related chemical groups	Example
Hexahedrites	H	>50	4.5–6.5	IIAB, IIG	Edmonton
Coarsest octahedrites	Ogg[a]	3.3–50	6.5–7.2	IIAB, IIG	
Coarse octahedrite	Og	1.3–3.3	6.5–8.5	IAB, IC, IIE, IIIAB, IIIE	Mayerthorpe
Medium octahedrite	Om	0.5–1.3	7.4–10.0	IAB, ID, IIE, IIIAB, IIIF	Iron Creek, Kinsella
Fine octahedrites	Of	0.2–0.5	7.8–13.0	IID, IIICD, IIIF, IVA	Millarville
Finest octahedrites	Off	<0.2	7.8–13.0	IIC, IIICD	
Plessitic octahedrites	Opl	<0.2, spindles	9.2–18.0	IIC, IIF	
Ataxites	D	–	>16	IIF, IVB	

Note: The table is based on Buchwald (1975).

[a]The g stands for the German grob, meaning coarse.

by assigning letters within each of the original groups (IA, IB, IIA, IIB, IIC, IIIA, IVB, etc.) as new groups were discovered; Iron Creek was classified as IIIA. Scott, Wasson, and Buchwald reported their neutron activation analysis of the Iron Creek meteorite: nickel (Ni; 7.72%); Ga (20.2 ppm); Ge (39.6 ppm); iridium (Ir; 3.3 ± 0.4 ppm).[35] Later, H. Haack and E.R.D. Scott (1993) reported INAA results for five more elements: cobalt (Co; 0.48%); chromium (Cr; 70 ppm); arsenic (As; 4.8 ppm); gold (Au; 0.75 ppm); rhenium (Re; 0.24 ppm); and a revised value for Ir (3.0 ppm).[36] D.J. Malvin, D. Wang, and J.T. Wasson provided a result for copper (Cu; 162 ppm).[37]

Not long after the meteorite arrived at Victoria College where it was examined and weighed by A.P. Coleman, and certainly before a cast was made of it, samples were sawed off and distributed. A 237 g specimen is held by the Field Museum of Natural History in Chicago; the U.S. National Museum in Washington, DC, has a 122 g piece; and there are progressively smaller samples at other institutions in Vienna; New York; London; Berlin; Ottawa; Tempe, Arizona; and Stockholm.

As Christopher Spratt has noted[38], the Iron Creek meteorite has always been stated to weigh about 386 pounds (175 kg), as first reported by Coleman[39], but evidently this weight was never verified until recently. After the meteorite was "repatriated" to Alberta, it was weighed by Don Taylor, Curator of Geology at the Provincial Museum of Alberta (now the Royal Alberta Museum), who determined it to be 145.25 kg (320 pounds).[40] Even allowing for the removal of the small samples listed above, there remains a discrepancy of some 66 pounds (30 kg) to be accounted for.[41] Spratt said there was undoubtedly an error in

Coleman's published weight.[42] The meteorite is presently on display in the Royal Alberta Museum's *Syncrude Gallery of Aboriginal Culture*. Ron Mussieux, the museum's recent Curator of Geology, said that although

> there have been consultations with the native people and with the scientific community concerning what should be done with this meteorite, [he does not know] what may occur to the Iron Creek meteorite in the future. [43]

Summary

The Iron Creek meteorite is a medium octahedrite, whose structure and softness suggest some annealing after a cosmic shock event, but it has no other unusual features of scientific interest. However, it is one of the most photogenic meteorites in the world, its shape bearing a striking resemblance to a human head in profile. Moreover, Iron Creek is of exceptional importance in the ethnohistory of Alberta and North American native peoples.[44]

3

The Kinsella Meteorite

History

In 1953, the Canadian meteoriticist Peter M. Millman lamented that

> the main mass, or biggest piece, has remained in Canada for only 13 of the
> [23 then known] Canadian meteorites....For at least five meteorites there
> seems to be no piece anywhere in Canada.[1]

Had he known about the Kinsella meteorite, he would have had to include it
in the latter category.

The American meteorite authority and collector Harvey H. Nininger
(1887–1986) believed that meteorites, although not abundant, were present
nearly everywhere and that ordinary people—farmers, townspeople, school
teachers, students—could find them if they were shown what to look for
and were offered money for any specimens found. Beginning in the 1920s,
Nininger developed and perfected a method or program for finding meteorites
that involved, among other things, public lectures and schoolroom talks that
included exhibiting examples of actual meteorites; interviewing fireball eyewitnesses and mapping their observations; talking to newspapers and mentioning
the offer of a cash reward for finders. Nininger's program was so successful
that his collection grew to be one of the largest and finest in the world, and his
methods have been widely copied to this day.

In May 1974, the University of California, Los Angeles (UCLA) issued a
press release offering to pay $100 for the first specimen of any new meteorite.[2]
During the following year, about 350 supposed meteorites were sent to UCLA,
of which seven were indeed new meteorites. Among the seven meteorites was
a 3.72 kg iron obtained from M.J. Malato of San Jose, California (Fig. 3.1). Mr.
Malato stated that it had been discovered on his brother-in-law Ollie Hallor's
farm midway between Minburn and Kinsella, Alberta. The meteorite had been
found in May 1946 about 20–30 cm deep during plowing by Mr. Hallor's son,
Ernest. Three decades later, the discovery location could only be estimated as
being 53°12′N, 111°26′W, with a probable error of ± 5 km.

Fig. 3.1. The Kinsella meteorite. Maximum length 21 cm. Thickness is only 1 cm in the cavity on lower left. Sawed face is perpendicular to the plane of photography on lower left. Note the irregular regmaglypts.

The discovery of the Kinsella meteorite was duly and belatedly reported in the *Meteoritical Bulletin* in 1978.[3]

The Kinsella meteorite was examined and weighed at the Geology Department, California State University, San Jose. Some traces of the outer heat-affected zone remain, and there is little corrosion. Kinsella weighed 3.72 kg intact. A slice weighing 84.5 g was retained at San Jose and was reportedly stolen. The main mass is in the UCLA collection; there are currently no samples in Alberta or Canada. A polished surface revealed the Widmanstätten structure of a medium octahedrite with a band width just over 1 mm (structural class Om). A small polished section showed the mineralogy typical of group IIIAB, the largest group of iron meteorites.[4]

Science

Two samples of Kinsella weighing 0.8 and 1.1 g were subjected to neutron activation analysis by Edward R.D. Scott, John T. Wasson and Richard W. Bild (University of California, Los Angeles). They were interested in whether Kinsella and Iron Creek, another IIIAB iron found only about 40 km away, were *paired meteorites*. Paired meteorites are those that were originally identified separately and thought to represent separate finds or falls but are shown to be the same on the basis of chemical data and petrographic characteristics. However,

Scott and his colleagues reported that the composition of Kinsella (nickel (Ni), 8.78%; cobalt (Co), 0.51%; gallium, 20.3 ppm; germanium, 42.0 ppm; iridium (Ir), 0.11 ppm; gold (Au), 1.20 ppm; arsenic (As), 9.0 ppm; tungsten, 0.50 ppm) was sufficiently different from that of Iron Creek for the two to be easily distinguished. Also, Kinsella was determined on the basis of its Ir/Ni ratio to be a IIIB iron, whereas Iron Creek is a IIIA.[5]

Alfred Kracher, John Willis, and John T. Wasson later compared the chemical composition of 57 iron meteorites including Kinsella and Millarville (Chapter 6).[6] In another paper (the tenth in the series) on the chemical classification of iron meteorites, Wasson's group reported that, in re-examining E.R.D. Scott's neutron activation analysis data, they had been able to detect the gamma-ray spectral signature of copper (Cu).[7] They were able to calculate the Cu content of over 100 iron meteorites including Kinsella (140 ppm) and Iron Creek (162 ppm), confirming that these two were not a paired fall. Later, H. Haack and E.R.D. Scott (1993) reported new instrumental NAA results: Co (0.51%); Cr (17 ppm); Ir (0.123 ppm); Au (1.2 ppm); As (8.6 ppm); and rhenium (<0.15 ppm).[8]

In our story of Alberta meteorites, Kinsella is chiefly of interest because it is "the one that got away;" there are no samples of it in Alberta or Canada.[9]

4

The Mayerthorpe Meteorite

History

In about 1946, a farmer near Mayerthorpe found a curious rock, unusually heavy for its size, on his land. He hacksawed off a piece, revealing a metal interior. The rock sat around for a while until the farmer, having lost interest in it, dumped it on a rock pile and forgot about it; later, much of the rock pile was incorporated into the foundation of a concrete pig pen. Years later (September 1964), another farmer, Mike Dmitroca, found a similar but bigger rock about one-quarter of a mile (0.5 km) farther east while cultivating a summer-fallowed field. He showed this second rock to his son Walter, a geophysicist, who promptly took it to the Department of Geology at the University of Alberta where it was identified as an 8.742 kg iron meteorite. Professor R.E. Folinsbee negotiated a deal with Walter Dmitroca; the University would purchase the meteorite from Mr. Dmitroca, senior, for $270 (to be increased to $540 if the meteorite should turn out to be a rare type—which it was not) plus a small polished and etched piece of the meteorite as a souvenir.[1]

The Mayerthorpe meteorite (now called Mayerthorpe #1) showed excellent regmaglypts, and its surface was slightly oxidized suggesting an old fall (Figs. 4.1 and 4.2).[2] The location of the fall was 10 miles (16 km) south of Rochfort Bridge, Alberta, in Section 17, Township 18, Range 23, west of the Fourth Meridian, corresponding to 53°47'N, 115°02'W.[3] The discovery and identification of this meteorite, plus the prospect of further payment, spurred a search of the old rock pile near the pig pen for the rock found, then dumped, years earlier. This rock (now called Mayerthorpe #2) was quickly recovered, identified as also being a meteorite, and sold by Walter Dmitroca to the University of Alberta for $120.[4] When recovered Mayerthorpe #2 weighed 3.865 kg. Subsequently, the university removed two small slices (20 g and 119 g, plus 15 g filings) parallel to the original surface hacksawed by the farmer. An examination of etched polished sections suggested Mayerthorpe #2 was an almost structureless meteorite about on the octahedrite-hexahedrite boundary (Fig. 4.3). A preliminary nickel (Ni) determination, on filings recovered from hacksawing, confirmed the low Ni content and suggested that it was a hexahedrite or even

Fig. 4.1. Mayerthorpe #1, the first fragment found of the Mayerthorpe meteorite. Note the well-developed regmaglypts and the oxidized (rusty) surface.

a nickel-poor ataxite.[5] When sold to the Geological Survey of Canada for $500, Mayerthorpe #2 weighed 3.666 kg.

When he reported the recovery of the second Mayerthorpe to Dr. E.L. Krinov, editor of the *Meteoritical Bulletin*, Folinsbee commented that:

> The evidence now suggests that Mayerthorpe was a meteor shower which occurred more than 18 years ago, with an elliptical fall pattern about two kilometers in length with a northeast-southwest elongation, with the larger individuals at the northeast apex of the ellipse. Since the area has been farmed for about 50 years and there is no record of any bright bolides being observed in the district it is likely that the fall occurred before farming began.[6]

Fig. 4.2. The reverse side of Mayerthorpe #1 shown in Fig. 4.1.

Science

Learning from the *Meteoritical Bulletin* about the recovery of Mayerthorpe #1 and #2, Hans Voshage (Max-Planck-Institut für Chemie, Mainz, Germany) wrote to Folinsbee in 1968 requesting a "10 to 20 g sample" for his ongoing program of measuring the cosmic-ray exposure ages of iron meteorites by the ^{41}K/^{40}K-^{4}He/^{21}Ne method, which is the mass-spectrometric analysis of the isotopes of potassium (K), neon (Ne), and helium (He) that are produced by cosmic rays.[7] As was his custom, Folinsbee promptly replied, sending Voshage a 15.4 g slice of Mayerthorpe #2 and asking only that "we would appreciate receiving a report of [your investigations] in due course."[8] Half a year later, when writing to request a sample of the Edmonton hexahedrite, Voshage briefly reported his results and an estimate of the cosmic-ray exposure age of Mayerthorpe: 510 ± 90 Ma.[9] He later published the full results of his investigation.[10]

The method employed by Voshage is based upon the fact that the ratios of three potassium isotopes (^{39}K/^{40}K) and (^{41}K/^{40}K) produced by cosmic rays are increasing with time owing to the radioactive decay of ^{40}K (half-life = 1.277 Ga). Potassium, which is extracted from an iron meteorite sample, exhibits a mass spectrum that depends on three quantities, namely: the cosmic-ray exposure age; the effective irradiation hardness of the sample, which determines the ratios of production rates of the three nuclides ^{39}K, ^{40}K, and ^{41}K; and the admixture of accessory potassium having the isotopic composition of terrestrial potassium (approximately 93% ^{39}K and 7% ^{41}K). Knowing two independent ratios, (^{39}K/^{40}K) and (^{41}K/^{40}K), is not enough to resolve the problem posed by the three unknown quantites. Hence, a third independent ratio of two rare gases, ^{4}He/^{21}Ne, must be measured, providing the information on the effective irradiation hardness (actually, the extent to which the sample was shielded from cosmic rays). Voshage measured the relative abundance of potassium isotopes as ^{39}K, 82.81%; ^{40}K, 4.15%;, and ^{41}K, 13.4%. The ^{4}He/^{21}Ne ratio was 294. He calculated a slightly revised exposure age of 495 ± 105 Ma. In the course of making the measurements described above, Voshage was also able to determine the ratio of lithium isotopes, ^{7}Li/^{6}Li (in Mayerthorpe, about 1.20). However, unlike the situation with potassium, he found no correlation between ^{7}Li/^{6}Li and ^{4}He/^{21}Ne, and he concluded that the ^{7}Li/^{6}Li ratio of cosmogenic lithium is essentially independent of the irradiation circumstances.[11]

Unfortunately, the exposure ages determined by the ^{41}K/^{40}K method do not always agree with the ages determined by various researchers using ratios of aluminum (Al), Ne, argon (Ar), and chlorine (Cl) isotopes, such as ^{26}Al/^{21}Ne, ^{39}Ar/^{38}Ar, and ^{36}Cl/^{36}Ar. In an effort to resolve this disparity, D. Aylmer and his colleagues measured the [26]Al content of Mayerthorpe (2.33 ± 0.23 dpm/kg; disintegrations per minute per kilogram of sample) and several other iron meteorites by accelerator mass spectrometry.[12] They concluded that exposure ages calculated from [26]Al/[21]Ne ratios cannot be calibrated so as to agree with both [41]K/[40]K ages and ages based on the shorter-lived nuclides [39]Ar and [36]Cl. They suggested that the problem may be the result of a 35% increase in cosmic-ray intensity during the last 10 Ma. Aylmer and co-workers also measured the [10]Be (beryllium) content of Mayerthorpe (3.50 ± 0.28 dpm/kg).

John Wasson (University of California, Los Angeles) was the first to provide a detailed description of the structure and composition of Mayerthorpe, in the fourth report of his lengthy investigation of iron meteorites (Table 4.1).[13] He placed it in chemical group I, which he defined as those iron meteorites with 190–520 ppm germanium (Ge), 6.4–8.6% Ni, 56–100 ppm gallium (Ga), and 0.6–5.5 ppm iridium (Ir). He measured these four elements in 87 irons, including Mayerthorpe, by the following methods: Ni by atomic absorption spectrometry of an aliquot of the dissoved sample; Ge and Ga by radiochemical neutron activation; and Ir by neutron activation. Wasson's mean values for Ni, Ge, Ga, and Ir in Mayerthorpe place it near mid-range for all four elements in the 87 meteorites. He put Mayerthorpe in structural class Og (coarse octahedrite), based upon his measured average Widmanstätten band width of 2.0 mm.

Byeon-Gak Choi, Xinwei Ouyang, and John Wasson later (1995) measured several more elements: chromium, cobalt, copper, arsenic, antimony, tungsten, rhenium, platinum, and gold plus the familiar Ni, Ge, Ga, and Ir in more than 100 IAB and IIICD iron meteorites, using instrumental neutron activation analysis (Table 4.1).[14] Once again, the elemental composition of Mayerthorpe proved unremarkable. The authors presented a revised classification scheme in which *magmatic* irons are believed, as before, to have formed in molten asteroid cores. They proposed an impact-melt model for *non-magmatic* irons in which individual IAB meteorites, such as Mayerthorpe, are formed in pools of melted material produced by [meteoritic] impacts into chondritic asteroidal surfaces. The lower melt temperatures produced by the impact process would result in less thorough mixing than in a molten core and this, combined with varying target and impactor characteristics, would account for the properties observed in non-magmatic irons. Furthermore, most melt pools would chill too rapidly to experience fractional crystallization. However, neither this model, nor any other, has been universally accepted and the origin of non-magmatic irons remains problematic.

John Wasson and G.W. Kallemeyn subsequently (2002) proposed a more detailed division of the non-magmatic irons in group IAB into five subgroups

Table 4.1. Minor and trace elements in the Mayerthorpe meteorite.

Method	Co (mg)	Ni (mg)	Cu (µg)	Cr (µg)	Ga (µg)	As (µg)	Ir (µg)	Ge (µg)	W (µg)	Au (µg)	Pt (µg)	Sb (ng)	Re (ng)	Reference
Atomic absorption spectrometry		71.9												Wasson 1970
RNAA					75.5			283						Wasson 1970
NAA						2.4								Wasson 1970
INAA	4.75	70.6	139	21	79.1	15.2	2.18	283	1.39	1.63	3.1	320	240	Choi et al. 1995
INAA	4.75	70.8	139	24	79	15.1	2.19	297	1.28	1.64	3.1	313	235	Wasson and Kallemeyn 2002

Note: Concentrations are given in units per gram of meteorite. Method abbreviations are as follows: NAA, neutron activation analysis; INAA, instrumental neutron activation analysis; RNAA, radiochemical neutron activation analysis.

and numerous "grouplets."[15] At the same time they provided a revised list of elemental abundances for two samples of Mayerthorpe and several hundred other IAB iron samples.

Summary

Mayerthorpe has been used in several important studies of the elemental composition and cosmic-ray exposure age of iron meteorites but has not been found to have any remarkable features.

5

The Millarville Meteorite

History

Farmers have found most of Alberta's meteorites. At 8:30 a.m. on Thursday, 28 April 1977, Chuck E. Hayward was plowing his field near Millarville, about 30 km southwest of Calgary when his machine struck what he, at first, thought was a large rock.[1] On closer inspection, however, Mr. Hayward, a veteran of several years of prospecting in the Yukon and Northwest Territories, recognized the heavy, triangular rock as an iron meteorite (Fig. 5.1). He contacted the University of Calgary's Department of Geology. Dr. Ed Klovan, then department head, and Dale Harvey, technical supervisor, travelled to Mr. Hayward's farm to examine it. The location of the find was measured with unusual accuracy: 900 feet (274 m) north, 1200 feet (366 m) west of the southeastern corner of Section 23, Township 21, Range 3, west of the Fifth Meridian.[2] Klovan offered Hayward $500 to borrow it for examination.

The 15.636 kg Millarville meteorite was found to be a relatively rare type of iron meteorite, and it soon came to the attention of John T. Wasson (University of California, Los Angeles) who travelled up to Calgary to investigate it. At a special ceremony at the University of Calgary in November 1977, the meteorite was unveiled (Figs. 5.2 and 5.3); Dr. Wasson gave a public lecture on meteorites; and Dr. Peter Millman, representing the National Research Council of Canada's Associate Committee on Meteorites (ACOM), presented Mr. Hayward with a commemorative scroll for recovering the Millarville meteorite.[3] The original plan was to keep the meteorite on permanent display in the Gallagher Library of Geology; however, when its value was appreciated, it was put in storage in the Earth Sciences Building at the University of Calgary.[4]

Science

The discovery of the Millarville meteorite was published in the *Meteoritical Bulletin* in 1979 by D.W. Harvey (University of Calgary) who reported it as a structurally anomalous group IVA iron.[5] Iron meteorites are grouped according to their chemical composition by comparing the concentrations of the trace elements gallium (Ga), germanium (Ge), and iridium (Ir) to the overall nickel (Ni)

Fig. 5.1. Mr. C.E. (Chuck) Hayward with the Millarville meteorite.

content to define distinct chemical groups. Other trace elements sometimes used to resolve groups include antimony, arsenic (As), chromium (Cr), cobalt (Co), copper (Cu), gold (Au), platinum (Pt), rhenium (Re), and tungsten (W). Members of group IVA are characterized by low Ga and very low Ge values. In 1980, Alfred Kracher, John Willis, and John T. Wasson (University of California, Los Angeles) used radiochemical neutron activation analysis and instrumental neutron activation analysis to measure the Ni, Ga, Ge, and Ir concentrations in 57 iron meteorites, including Millarville (Table 5.1).[6] They also reported that Millarville has a thoroughly reheated structure. Two decades later, significant improvements in the quality of neutron activation analyses prompted J.T. Wasson and J.W. Richardson to restudy many of the meteorites in the earlier report. Because the range of Au values is several times larger than that of Ni, they chose to compare the concentrations of Ni, Co, and several trace elements to the Au content. Besides giving revised values for Ni, Ga, Ge, and Ir levels, the new study also included measurements of Cr, Co, Cu, As, W, Re, Pt, and Au.[7]

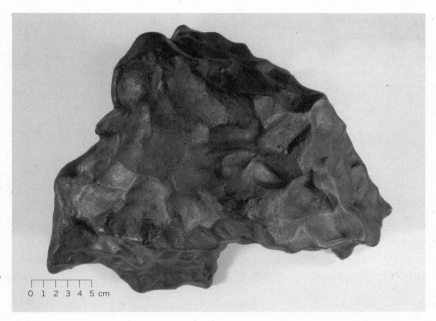

Fig. 5.2. A cast of the Millarville meteorite. Note the well-developed regmaglypts.

Fig. 5.3. Another view of the cast of the Millarville meteorite.

Table **5.1.** Minor and trace elements in the Millarville meteorite.

Method	Co (mg)	Ni (mg)	Cu (µg)	Cr (µg)	Ga (µg)	As (µg)	Ir (µg)	Ge (µg)	W (µg)	Au (µg)	Re (ng)	Reference
INAA	4.17	93.0	133	10	2.19	11.3	1.02		0.44	2.43	150	Wasson and Richardson 2001
RNAA								0.144				Wasson and Richardson 2001
INAA/RNAA		95.7			2.38		0.98	0.144				Kracher et al. 1980

Note: Concentrations are given in units per gram of meteorite. Method abbreviations are as follows: INAA, instrumental neutron activation analysis; RNAA, radionuclide neutron activation analysis.

Yet another way that iron meteorites can be classified is into magmatic irons and non-magmatic irons (see Chapter 5). Most iron meteorites, including groups IIAB (e.g., Edmonton), IIIAB (e.g., Iron Creek and Kinsella), and IVA (e.g., Millarville), are magmatic irons; groups IAB (e.g., Mayerthorpe), IIE, and IIICD are non-magmatic.

Two

Stony Meteorites

6

The Belly River Meteorite

History

Lincoln LaPaz, Director of the Institute of Meteoritics (IOM; University of New Mexico, Albuquerque), provided the first riveting account of the discovery of the Belly River meteorite:

> In the winter of 1943–4, a R.C.A.F. plane made a forced landing on the eastern side of and near to the Belly River in southern Alberta, Canada, at a point approximately in longitude W. 113°00′ and latitude N. 49°30′. Chief Photographic Officer B. Wettlaufer was one of those aboard the plane who set out on foot to scout out the area in which the plane had been forced to land. In the course of his reconnaissance, he chanced upon a dark, reddish-brown rock exhibiting such a remarkably pitted and sculptured surface that he picked it up and lugged it all the way back to the plane, altho [sic] it weighed in excess of 17 lb [8 kg]. And his return hike was made in a near blizzard with winds of such force that he was sometimes thrown off his feet....[1]

The reality was somewhat less dramatic. Sixty years after finding his meteorite and half a century after LaPaz's published account, Boyd Wettlaufer (Fig.6.1) told the author what really happened:

> I was an instructor [in aerial photography at the RCAF training station at Pierce, near Fort Macleod, Alberta, during the 1943–1944 period]. My job was to train student pilots in photography. I would take up four students at a time in an Anson aircraft and assign them a number of targets to photograph—a farmer's barn, an elevator, a church steeple, etc. so I flew over every part of the Fort Macleod area. During one of these flights I spotted some teepee rings on the east bank of an old meander of the Oldman River and that weekend [my wife] Dorothy and I went to see if they were of native or RCMP [Royal Canadian Mounted Police] origin.
> There was a big wind blowing, big for even Fort Macleod, which was noted for its winds. So my wife stayed in the car and left the looking up to

me. We were parked on the road [Highway 3] just north of the boundary line of the [Peigan] reserve a few miles south of Macleod. I had about 500 metres of a walk; checked out the teepee rings and started back across the farmer's plowed field. The farmer had stones off the field piled in a couple of piles. Since it was customary to check stone piles for Indian stone hammers, I checked them out—no stone hammers, but one very interesting stone—it had a rusty iron appearance and had pockmarks in its surface, as if it had been subjected to great heat. Also it was quite heavy for its size (about 25 to 30 cm long and about 15 cm thick). The only thing that fitted that description was a meteorite—so naturally I carried it back to the car. I was now facing the wind and it was quite a struggle. I fought the wind for about a hundred yards [90 m] and said to myself: "to heck with this" and threw the meteorite down and started back to the car. Then I had second thoughts and said to myself: "no, that's a scientific specimen. I must pick it up." So back I went and brought it to the car. Dorothy was quite fascinated by it and agreed that it had to be a meteorite.

We had it with us from then on. I had several more postings—Portage la Prairie, Montreal, Ottawa and Edmonton where I was at the end of the war. I left the service and was able to go to New Mexico in 1946; I graduated [from the University of New Mexico] as an archaeologist in 1949. During this time I found the Institute of Meteoritics and sold the meteorite to Dr. LaPaz with the proviso that he share it with Canada. I went on to take graduate studies at U.N.M. and took an [archaeological] expedition up to Alberta in 1949 at Head-Smashed-In Buffalo Jump near Fort Macleod.[2] In 1965, I was going through Alberta and rechecked the old stone piles—no more meteorites.[3]

Boyd Wettlaufer believes that Lincoln LaPaz was writing from memory when, six or seven years after he purchased the meteorite, he published the first description of the Belly River meteorite in a preliminary note.[4] Consequently, not just the discovery details, but also the reported geographical location of the find are in error.[5] With the aid of topographical maps provided by the author Wettlaufer was able, in 2004, to retrace the route of his trek and reconstruct the events leading to his discovery of the meteorite 60 years earlier.[6] He identified the place where he parked his car that day, the terrain he followed, and the farmer's field where the rock piles were situated. The true location of where the Belly River meteorite was found is on the eastern side of the Oldman River (so the *Belly* River meteorite is really

Fig. 6.1. Chief Photographic Officer Boyd Wettlaufer (RCAF), circa 1943–1944, discoverer of the Belly River meteorite.

the *Oldman* River meteorite) at 49°39′44″N, 113°29′22″W. In the system of land descriptions on the Canadian prairies, the field is identified as NW 1/4, Section 20, Township 8, Range 26, west of the Fourth Meridian. This location is 40 km northwest of the location given by LaPaz.[7] Although the name Belly River is inappropriate, it has been in use since 1953 and was accepted by the *Meteoritical Bulletin*, so it is too late to change the name. This is unfortunate because another meteorite really was found near the Belly River in 2004 and had to be provisionally named Belly River Buttes (Chapter 7) to distinguish it from the misnamed Belly River meteorite. Meteoriticists are always interested in meteorites that are found within a limited geographical area in case they are actually two members of a single fall. However, in this instance, Belly River (an H6 (S4) chondrite) is clearly not related to the newly found Belly River Buttes (an L6 (S3) chondrite).[8] Much of the outer surface of the Belly River meteorite is covered by shallow oval pits or depressions called regmaglypts.

Science

Mineralogy and Textures

Lincoln LaPaz, Director of the IOM, University of New Mexico, Albuquerque, provided the first physical details of the Belly River meteorite, describing it as a "roughly triangular prism with rounded ends, covered by a dark-red to a reddish brown fusion crust (Figs. 6.2 and 6.3)."[9] It had a mass of 7900 g and a density of 3.6 g/cm^3. He classified it as a veined chondrite iron, C(v).[10] The most interesting feature of the meteorite, according to LaPaz, is

> its remarkable system of veins as revealed on cut and polished surfaces. These range in width from almost invisible sinuous hairlines, which, in spite of their narrowness, extend in some cases more than 50 mm, and under moderate power [magnification], are seen to be made up of numerous, still finer metallic threads, to elongate, irregular inclusions with maximum widths of 3 mm, consisting in some cases of solid metal, in others of a filamentary metallic gauze.[11]

In late 1960, after the fall of the Bruderheim meteorite (see Chapter 8) LaPaz wrote to R.E. Folinsbee, Chairman of the Department of Geology, University of Alberta, offering to exchange an equal-sized sample of the Belly River meteorite for a specimen of Bruderheim "in the 1–3 kg range."[12] Folinsbee accepted this offer with alacrity, because the University of Alberta had over 300 kg of Bruderheim but not a single piece of Belly River in its collection. In the initial swap, the IOM sent the University of Alberta two slices of Belly River weighing 1.73 kg and 0.96 kg, respectively; ironically, the 1.82 kg uncut Bruderheim specimen the IOM received was, in LaPaz's words, "too fine a museum specimen to be cut up even for such valuable scientific work as that being done...."[13] So, to complete the exchange, Folinsbee sent two additional, presumably more expendable Bruderheim specimens. Less than three months later, Folinsbee traded the larger Belly River specimen to the Geological Survey of Canada for a 2.18 kg specimen of the Abee meteorite (see Chapter 15).[14]

The mineralogical and chemical composition of the Belly River meteorite were investigated and compared with that of three other chondrites by Brian Mason and H.B. Wiik of the American Museum of Natural History.[15] They determined the total iron content to be 25.89% and placed the meteorite in the high-iron (H) group of Nobel Prize winner Harold Urey and H. Craig's classification scheme.[16] Some of the metallic nickel-iron occurs in large chunks (up to 1 cm in diameter), but most appears as smaller flakes scattered throughout the dark matrix and in the veins previously described. Iron sulphide is

Fig. 6.2. The Belly River meteorite shows well-developed regmaglypts. Note the flat end where a sample has been sawed off.

present as the mineral *troilite*. The most common mineral present is olivine, an iron-magnesium silicate. Typical olivine is what is known as a *solid solution series*, that is, it may assume compositions ranging between two pure "end-members," one iron-rich (fayalite, Fa) and one magnesium-rich (forsterite, Fo). As in all common meteorites, the olivine in Belly River has a majority of the magnesium-rich end-member[17] and its composition may be expressed as Fa_{20}, meaning that it contains 20% of the fayalite molecule, 80% of the forsterite molecule. Pyroxenes are more complex solid solutions, with three end-members (rich in iron, magnesium, and calcium) required to describe adequately most common varieties; the pyroxene *bronzite* is the second-most abundant mineral in Belly River. Indeed, Belly River and other H chondrites were formerly known as *olivine-bronzite* chondrites (Fig. 6.6).

Chondrites are also classified on a scale ranging from H3 to H6 according to their petrologic type (there are no H1 or H2 chondrites); that is, on the extent to which they have been modified by heat prior to arriving on Earth. This metamorphism is familiar to us on Earth as the process that changes limestone into marble or shale into slate. Petrologic type H3 corresponds to the least altered state, whereas type H6 chondrites have undergone such intense heating and

Fig. 6.3. A cut and roughly polished slab of the Belly River meteorite. The cracks in the dark gray matrix are caused by weathering, suggesting that the meteorite lay on (or in) the ground for years before being found.

recrystallization that chondrules may be almost completely obliterated. As an H6 chondrite, the Belly River meteorite has undergone significant thermal metamorphism.

Chondrites, the most common type of stony meteorites, also show signs of alteration by yet another process called *shock metamorphism*.[18] Shock metamorphism results from the violent collision between a parent body and another asteroid or meteorite. Certain conspicuous features, such as the shock veins seen in the Belly River meteorite (Fig. 6.4), have long been recognized but, only recently, have the full range of shock-induced features been catalogued and organized into a scale of progressive levels, or stages, of shock. Dieter Stöffler and his colleagues (University of Hawaii) defined six stages: S1 (unshocked); S2 (very weakly shocked); S3 (weakly shocked); S4 (moderately shocked); S5

Fig. 6.4. A polished thin section of the Belly River meteorite seen in transmitted plane-polarized light. Opaque shock veins can be seen running diagonally across the field. (35× magnification). Field of view 3.3 mm. For colour image see p. 287.

(strongly shocked); and S6 (very strongly shocked).[19] Each stage is characterized by distinctive, visible (with a polarizing microscope) physical changes in the minerals olivine, pyroxene, and plagioclase. Furthermore, by using experimental shock pressure data, Stöffler and his co-workers were able to calibrate their scale; the S3/S4 transition, for example, occurs at a pressure of 15–20 GPa (giga-Pascals; normal atmospheric pressure is about 100 kiloPascals, or 0.0001 GPa). Determining the shock stages of meteorites can thus provide much information about the collisional and geological history of their asteroidal parental bodies.

Meteoritic bombardment has been pervasive throughout the solar system; every inner planet, satellite, asteroid, and comet nucleus that has been visited by spacecraft shows signs of extensive impact cratering. This bombardment creates shock waves in both target and impactor that first compress, then decompress the material; the shock wave energy is converted to heat, heating the material to the so-called *post-shock temperature*. This temperature increases with increasing shock pressure and may cause melting or vaporization at very high shock pressures.

Fig. 6.5. A polished thin section of the Belly River meteorite seen in transmitted cross-polarized light. The large green mineral is olivine; note the shock-induced fracturing (42× magnification). Field of view 2.7 mm. For colour image see p. 287.

Applying their new shock stage analysis to the Belly River meteorite, Stöffler and his colleagues assigned it a S4 shock stage because of the presence of opaque shock veins, which form at relatively low shock levels, and pervasive melt veins and melt pockets, which occur at somewhat higher shock levels and coexist with opaque shock veins. All three features—shock veins, melt veins, and melt pockets—were formed by the shock-induced, localized simultaneous melting of silicate, sulphide, and metal grains. Mineral grains that escaped melting nevertheless show distinctive and characteristic shock effects such as fracturing, deformation and recrystallization (Figs. 6.5 and 6.6).

Graham C. Wilson (Turnstone Geological Services, Campbellford, Ontario) studied a polished thin section of Belly River microscopically in both reflected and transmitted light concurrently with his petrographic and mineralogical examination of the Belly River Buttes meteorite.[20] Wilson described Belly River as consisting predominantly (54%) of a very fine-grained matrix of pyroxene with lesser amounts of feldspar and olivine grains, all 3–15 μm in diameter and variably iron stained. Scattered throughout the matrix are frequent (8%) highly

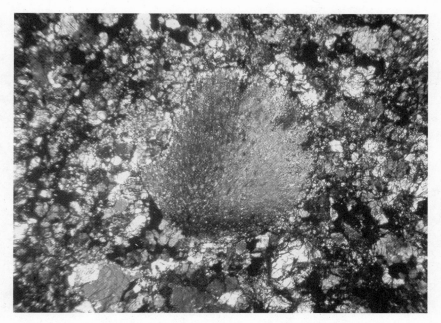

Fig. 6.6. A radial pyroxene chondrule in the Belly River meteorite (cross-polarized light). Recrystallization has partially obliterated the margin of the chondrule. (32× magnification). Field of view 3.6 mm. For colour image see p. 288.

reflective, grey-white flakes of nickel-iron alloy; these are mostly kamacite with possible traces of taenite, which Wilson described as "pale ghostly traces on the margin of host kamacite," and a maximum grain size of 1000 × 700 μm. Troilite is present (5%) as brownish-yellow grains much duller than the alloy flakes. Among minor minerals present are red translucent grains and veinlets of *goethite* and grey *chromite*.

Also present are several types of chondrules. *Porphyritic* olivine chondrules (constituting 8% of the meteorite) have a maximum size of 1.2 × 1.0 mm and consist of discrete angular olivine crystals, up to 450 × 200 μm in section, in a very fine-grained (<3 μm) matrix; these chondrules and chondrule fragments, or relicts, may be hard to distinguish from the matrix or groundmass of the meteorite. Less common are barred olivine (2%) and barred orthopyroxene (2%) chondrules. Rarer still (1%), but notably well defined against the matrix, are excentroradial pyroxene chondrules (Fig. 6.6) and large, very fine-grained chondrules (1%) possibly composed of plagioclase. Wilson also found scattered large (up to 1600 × 600 μm) angular grains of olivine (7%), and coarse (~600 × 600 μm) colourless grains of orthopyroxene (4%) and plagioclase feldspar (3%). Under a petrographic microscope, both the olivine chondrules and the coarse olivine crystals display signs of strain, such as planar fractures (straight, parallel cracks) and mosaic structure (spotty, irregular changes in colour as an olivine grain is rotated in polarized light); these signs, plus evidence of strain in the coarse orthopyroxene and plagioclase feldspar grains, as well as the presence of inter-connected shock veins, are indicative of an S4 shock stage.

Summary
The first H chondrite to be found in Alberta, Belly River is chiefly notable for its opaque shock veins, melt veins, and melt pockets. It is a tribute to the perspi-cacity of its discoverer, Boyd Wettlaufer, whose later research established the archaeological significance of the Head-Smashed-In Buffalo Jump, not far from where he found his meteorite.

History

Alberta's fifteenth meteorite was found in 1992 by Mr. Gerald Goldenbeld while baling straw on his farm south of Fort Macleod on the west bank of the Belly River, opposite the Belly River Buttes.[1] He spotted an unusual black-and-rust-coloured stone in his field and stopped his tractor to pick it up. He kept the heavy fist-sized rock, which he thought might be a meteorite, in his house, showing it to people from time to time, for several years. In early 2004, he got interested in confirming its identity as a meteorite and searched the Internet for pictures of meteorites and other information. Goldenbeld then sent an email with an attached photograph to Dr. Alan Hildebrand at the Department of Geology and Geophysics at the University of Calgary.[2] Hildebrand recognized the specimen as a genuine meteorite as soon as he saw the photograph and made arrangements to acquire it for the University of Calgary. The 1.5 kg provisionally named Belly River Buttes meteorite is moderately weathered but still retains its fusion crust. The meteorite has apparently been broken in two along a glassy shock vein, likely by farm equipment; so there may still be another part of the meteorite lying in the field, awaiting recovery.

Science

A small piece of the meteorite was cut off and a polished thin section, 20×15 mm in maximum dimensions and 35 μm thick, was prepared. The section was then coated with a thin film of carbon to make the surface electrically conductive, in preparation for electron microprobe analysis, which is performed by detecting and counting the fluorescent X-rays produced by bombarding a sample with high-voltage electrons from an instrument similar to an electron microscope.

Exhaustive analysis of olivine was undertaken by Hildebrand's group at the University of Calgary.[3] Hildebrand found that the olivine in the Belly River Buttes meteorite averages 24.7% fayalite (abbreviated $Fa_{24.7}$), the iron-rich variety. He also reported finding clinopyroxene, indicating that Belly River Buttes, like Ferintosh (Chapter 9), is a two-pyroxene meteorite. Following this analysis, Hildebrand loaned the polished thin section to Graham C. Wilson (Turnstone

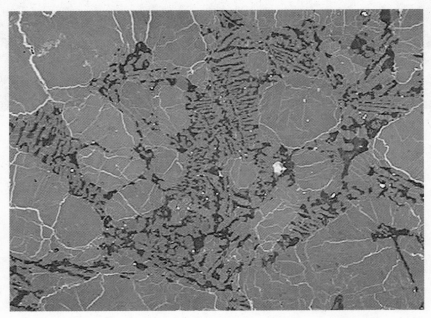

Fig. 7.1. An unusual array of small (e.g., 30 × 5 µm) crystallites of olivine in glass, perhaps a small, incipient melt pocket interstitial to much larger (100–400 µm) tabular olivine crystals in the Belly River Buttes meteorite. Nominal magnification 100×. Plane-polarized reflected light.

Geological Services, Campbellford, Ontario) who conducted a comparative petrographic and mineralogical study of it and the Belly River meteorite.[4]

Wilson received the Belly River Buttes sample in June 2004 and cleaned it, removing the carbon coating remaining from the electron microprobe analysis, before beginning his microscopic examination using both reflected and transmitted light. Differences between the two meteorites were quite apparent; for a start, the mainly olivine and orthopyroxene matrix comprises more (62%) of Belly River Buttes and is somewhat coarser (average grain size, 5–20 µm) and more strongly iron stained than that of Belly River. Highly reflective grains of kamacite, so prominent (8%) in Belly River, are less conspicuous in Belly River Buttes (only 3%). However, Belly River Buttes contains more chondrules; these are mainly porphyritic olivine chondrules (15%) up to 2.2 × 1.8 mm in size, with olivine *phenocrysts* up to 400 × 400 µm in a matrix variably composed of coarse olivine and fine-grained material (Fig. 7.1). There are also lesser numbers of barred olivine chondrules (2%) with a maximum size of at least 800 × 700 µm; excentroradial pyroxene chondrules (2%) up to 1.2 × 1.0 mm; and very fine-grained chondrules (1%) up to 1.4 × 1.2 mm in section.

Troilite in Belly River Buttes is present as brownish-yellow masses up to at least 360×250 μm, in similar quantity (7%) as Belly River but less shocked (smaller areas of mosaic structure) than in the latter. There are also occasional (1%) coarse grains (maximum grain size, 330×250 μm) of orthopyroxene and traces of smaller (up to 120×60 μm) plagioclase feldspar. Among other minerals present are traces of goethite and chromite.

The $Fa_{24.7}$ olivine of the Belly River Buttes meteorite indicates that it is an L chondrite, while

> the poor to moderate definition of most chondrule margins and the presence of relatively coarse feldspar are consistent with an L6 stone. The moderately strained and deformed olivine and limited presence of shock veins suggest an S3 shock state.[5]

An interesting feature of the Belly River Buttes meteorite is the presence, in the up to 150 μm wide dark veins, of 2–10 μm spheroids of troilite; these show up especially well in reflected light.

Belly River Buttes is an ordinary L6 chondrite, like Bruderheim, Ferintosh, Peace River, and Vulcan. It is quite unremarkable, as meteorites go, but there is a story behind its name. It had to be provisionally named Belly River Buttes to avoid confusion with the misnamed Belly River meteorite, which was actually found near the Oldman River (see Chapter 6).

8

The Bruderheim Meteorite

History

No meteorite has ever fallen on Canada as propitiously as the Bruderheim meteorite. It fell in late winter in the midst of an agricultural area dotted with several small communities and numerous farm houses. Fifty kilometres to the southwest was the city of Edmonton, home to the University of Alberta, the Alberta Research Council, the Edmonton Centre of the Royal Astronomical Society of Canada, Canada's first public planetarium—the Queen Elizabeth II Planetarium, and a full complement of news media and communications facilities.[1] Also to the southwest, and nearer still, in the town of Fort Saskatchewan was the Sherritt Gordon Mines Limited nickel refinery, with its well-equipped metallurgical laboratory.[2] All of these would play a vital role in the recovery of the meteorite.

At the beginning of March 1960, the Earth was just two months past perihelion—the point in its orbit closest to the Sun, at which time its orbital velocity is the highest—and it was still rushing along at about 30 km/s. Even so, it was being rapidly overtaken by a chunk of asteroid travelling at more than 40 km/s (Fig. 8.1). This meteoroid won the race with our planet at 1:06 a.m. MST on Friday, 4 March 1960, only to slam into the upper atmosphere with a still impressive net velocity of over 13 km/s (~50,000 km/h). The intense friction caused by air resistance, even at this high altitude, almost immediately heated the meteoroid and the air surrounding it to brilliant incandescence. The Bruderheim bolide was probably first observed by Alexis Simon, a resident of the Paul's Band Indian Reserve at Duffield, just as it entered the atmosphere. His account is as follows:

> On the night of Friday, March 4th, 1960, I happened to be outside of my home at midnight when I saw a large meteorite in the north-westerly direction from Duffield. It lighted up the sky as it passed swiftly in a north-easterly direction, giving off what appeared to be flashes of fire.[3]

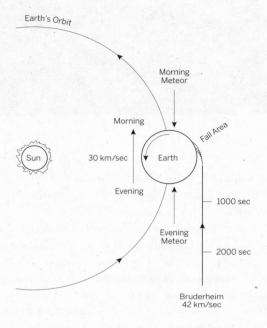

Fig. 8.1. A schematic diagram showing how the Bruderheim meteorite approached the Earth.

At 1:06 a.m., the late show on CFRN television had just finished, a factor adding substantially to the number of eyewitness accounts. There was considerable patchy overcast in the Edmonton area, and the meteor cameras were not operating at the Newbrook Observatory (see Chapter 16); however, the all-sky camera was on and recorded the general brightening caused by the bolide. Mrs. P.I.B. Wood of Carvel (Fig. 8.2) observed the fireball from the point of detonation to disappearance, as did a number of observers from the slightly cloudy city of Edmonton. Some of the most accurate observations were made from Edmonton by D.B. Russell, a student at the University of Alberta, and from Beverly by M. Reis at the Texaco Oil Refinery. Cross sightings were made by S.E.J. Mitchell of Clyde and by a number of observers near Egremont and in Fort Saskatchewan. Some 160 km east of the fall area, Mr. and Mrs. A.C. Butz of Dewberry observed the flash and "about two to three minutes after light, a thundering noise was heard, windows rattled...."[4]

The Sherritt Gordon nickel refinery's employee newsletter, the *Nickelodeon*, recorded the events of that early morning as experienced at the plant and in the town of Fort Saskatchewan:

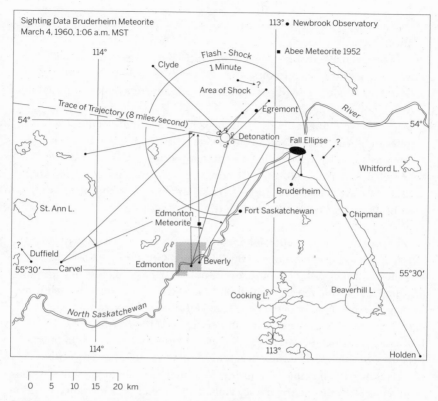

Fig. 8.2. Visual sightlines of bolide observers, point of detonation, and fall area for the Bruderheim meteorite.

The blinding, bluish, flash and the unseasonal thunder-like crash, followed by the long, loud rumble, heralded the arrival of the meteorite, which exploded almost immediately above our community....

For those of us who witnessed this eery [*sic*] and spectacular celestial display, it was an experience never to be forgotten. Many stood petrified as windows vibrated in their frames and dishes rattled on the shelves. The night constable patrolling the business area of the Town, stood with blinded eyes as the store windows testified to the severity of the shock as they vibrated for almost a full half minute. An employee of the Company, working in the lonely area of the Tailings Pond, shaken by the experience, headed back to "civilization" and the company of his fellow workers. In scores of homes residents were awakened from their sleep and wandered around, curious and bewildered. Frightened children cried and sought the comfort of their parents' arms. This was the scene in our Town early that morning, although some slept on unaware that history was being made in the heavens.

Local, reliable reports from eye witnesses indicate that the fireball streaked across our sky towards the east and exploded with a blinding flash with the brilliance of daylight almost immediately above, although slightly to the north, of the "Fort."

Upon exploding the body continued on in three main sections, one straight ahead and one on either side. At 1:30 a.m. the Dominion Meteorological Bureau at the Edmonton Airport was contacted and a detailed report was registered. The time lapse between the explosion and the first crash was given as forty-five to sixty seconds and the loud rumbling continuing for a period of twenty to thirty seconds.[5]

The press, radio, and television media were very cooperative and highly effective in helping Earl Milton, of the Edmonton Centre of the Royal Astronomical Society of Canada, alert the public to the possibility that a mete-orite had fallen from the exploding bolide. Ian McLennan (who would shortly become the first director of the new Queen Elizabeth II Planetarium) and other staff at CFRN-TV were especially helpful. Nick Broda, a farmer in the Bruderheim district, recovered the first piece of meteorite from his barnyard later that same day, Friday, March 4. It was brought to the Sherritt Gordon nickel refinery by an employee and later identified as a meteorite; this news was soon reported by the media. Stanley Walker, the refinery's safety supervisor, had already talked to the midnight shift workers as they came off duty when he heard the report about the farmer who had brought in a black rock for identifi-cation. He picks up the story:

> As soon as I heard "black rock," I knew it was part of the meteorite. Taking a young man from our local rock club, Tyrone Balacko [Fig. 8.3], Saturday morning at daybreak we were knocking at the farmer [Nick Broda]'s door asking for permission to hunt on his property. It was 10 °F [-12 °C] and there was six inches [15 cm] of snow....
>
> On cultivated land, the snow was only 3 inches [8 cm] deep and small pieces of the meteorite spewed up dirt on top of the snow and were quite conspicuous at 100 yards [91 m].[6]

Larger fragments, falling at a terminal velocity of about 300 km/h, hit the frozen ground, leaving shallow craters and rebounding with a shower of dirt onto the surface of the snow (Figs. 8.4 and 8.5). Stan Walker and Ty Balacko

Fig. 8.3. Ty Balacko (left) and Stan Walker with some of their finds of the Bruderheim meteorite; the two largest pieces weigh about 20–30 kg.

travelled the countryside around Bruderheim all day Saturday and Sunday. They ended up collecting over a dozen meteorites with a total mass of some 70 kg. As they found pieces, Walker and Balacko carefully marked the locations of their finds on a map and took photographs of the impact sites. Next day, Monday, there was a heavy snowfall that prevented any further searching until near the end of March. On Tuesday, Stan Walker hosted a meeting at his home:

> comprised of the Department of Geology from the University of Alberta, representatives from radio, TV and newspaper, the Royal Astronomical Society of Canada, and my immediate boss. They all wanted a piece of the meteorite. It was agreed by the University and ourselves that: [7]
>
> 1. The largest piece would go to the Edmonton planetarium for permanent display.[8]
> 2. One piece would be donated to the Fort Saskatchewan High School.[9]
> 3. Two pieces each to my partner and myself.

Fig. 8.4. Stan Walker and Ty Balacko's photograph of the plunge pit formed by the Bruderheim individual B-12 (3.03 kg), which fell in a field of swathed wheat.

All of this would occur after the university had completed their studies. We had made a map of our findings on each quarter section of land. We had also taken some pictures. For our pictures, our map, and our material the University paid us $500.[10]

Their modesty ("made a map, took some pictures") aside, Walker and Balacko's meticulous documenting of their findings earned high praise. Dr. R.E. Folinsbee, Head of the university's Department of Geology, said that they "have done an admirable job of mapping and collecting..."[11] Earl Milton, a past president of the Edmonton Centre of the Royal Astronomical Society of Canada (RASC), commended them for "photographing the fragments *in situ* and in making a scale drawing of the particle distribution over the area of the fall."[12]

Ironically, all these events—from the spectacular fireball through the meteorite search by Walker and Balacko—happened while Folinsbee was out of the country; he was in New York City, attending a U.S. National Academy of Sciences symposium on the ages of rocks.[13] When he returned on Sunday, March 6, Folinsbee found that, in his absence, the Edmonton Centre of the RASC had very efficiently handled the early collection of fireball sighting reports and the organization of search and recovery efforts; he assumed

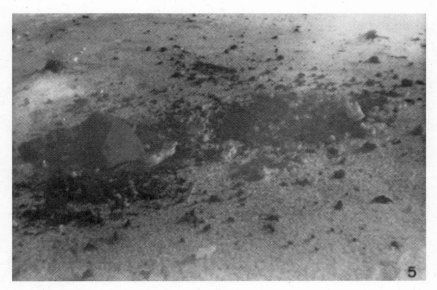

Fig. 8.5. Bruderheim B-10 (3.13 kg), which fell on fallow ground.

responsibility for these activities. One of his priorities was obtaining fund-
ing from the University of Alberta for meteorite purchases; W.H. Johns, the
university's president, made a special fund available for acquisition of the
Bruderheim meteorite.[14] Folinsbee was also concerned with preventing the
dispersal of the Bruderheim fall to competing agencies and private individuals
and the attendant increase in meteorite prices. In the ensuing negotiations, co-
ordinated by the RASC,[15] the Geological Survey of Canada (GSC) agreed to recognize the primacy of the University of Alberta as sole purchaser of Bruderheim
fragments; in exchange, the university agreed to provide (at cost) the GSC with
one large specimen and a few smaller ones for the national collection.[16] A
report on the fall of the Bruderheim meteorite shower was quickly sent to E.L.
Krinov in Moscow for publication in the *Meteoritical Bulletin.*[17]

Another of Folinsbee's pressing concerns was establishing legal ownership
of the meteorites. At first, he had accepted the legal opinion "given me by an
eminent Edmonton Q.C. [Queen's Counsel]...[who] maintained that meteorites would be the property of the finder."[18] Then the farmer, on whose land
the large (30.7 kg) specimen that Folinsbee intended to send to the National
Meteorite Collection in Ottawa had fallen, claimed ownership. Now Folinsbee
found out that, while meteors are ownerless objects, "in the opinion of the Law
Department at the University, a case can be made out that once they land they

become the property of the user of that particular piece of land."[19] He decided to settle out of court, paid the landowner $201 (in addition to the $175 he had already paid the original vendor). Prudently, "title to the Walker-Balacko collection [which also had already been paid for once] was ensured by settling with the farmers on whose land these fragments were collected...."[20]

By now it was spring, and a concerted effort, driven by a sense of urgency, was made to recover any remaining Bruderheim meteorites. It was a race against time: the searchers had to wait for the black fragments to become exposed on top of the melting snow, but wait too long and the fragments could not be seen and collected against the black soil of the fields and "undoubtedly many thousands of them were ploughed under."[21] It was soon concluded that local residents could search the fields more efficiently while preparing for seeding than searchers from the university, and the majority of specimens were purchased from Bruderheim farmers. University collecting concentrated instead on the frozen surface of the nearby North Saskatchewan River. Hundreds of grit- and pebble-sized fragments were collected off the river ice, "with collection continuing until a day or so before breakup—the last collection we made roped together, and with a long pole for support lest the ice give way."[22]

Folinsbee also systematically tracked down and purchased all worthwhile specimens from farmers and collectors. He paid farmer Fred Krys five dollars a pound ($11/kg) for one "exceptional museum specimen completely coated with a fusion crust and with no chips out of it."[23] On the other hand, he purchased about 9 kg of weathered, broken and chipped fragments from Mr. Krys's neighbour, Nick Broda, for $25. Mr. Fred Alexandruk, principal of Walker School, Bruderheim, sold his 17.3 kg specimen (B-37) to the university for $60.[24] One large individual, found on the day of the fall by Bruderheim area farmers, had been broken into several pieces and distributed to families and friends; Folinsbee managed to recover 25 kg of an estimated 30 kg original mass. Because this individual (B-74) was of little value for museum purposes, and had been recovered within a day of its fall, it was further divided into samples for distribution to researchers. Folinsbee traced another errant 2 kg fragment to a collector in Toronto. Several months later, Jack Grant, from the Meanook Geomagnetic Observatory (see Chapter 16), told Folinsbee about a Bruderheim fragment he, Earl Milton, and Ian McLennan had found on the farm of a Mr. Onushko. "It was," Grant said, "found in a grove of aspen trees and had broken a live, but frozen, branch off one of them in its fall."[25] By summer, the University of Alberta had amassed some 300 kg of material from the Bruderheim fall, quadrupling Walker and Balacko's original haul.[26] In 1989, a "hitherto unrecorded fragment of the Bruderheim meteorite," weighing 3.65 kg and in excellent condition with an almost complete fusion crust, was reported in Victoria, British Columbia.[27] As recently as 2005, the University of Alberta Meteorite Collection received a donation of a 1.2 kg piece of the Bruderheim meteorite from Mrs. Ruby Beaman of Fort Saskatchewan.

Science

Fall Phenomena and Astronomical Interpretation

An ellipse of fall, 5.6 km long and 3.6 km wide, was plotted with the aid of the Walker-Balacko map, aerial photographs, and other sources of information (Fig. 8.6). Two-thirds of the material recovered from the Bruderheim fall was in the form of large (>4 kg) meteorites; there were 12 such individuals, ranging in weight from 4.56 to 31.36 kg, with an aggregate mass of 208 kg.[28]

Even as the recovery of Bruderheim fragments continued through the spring and summer of 1960, Folinsbee embarked on an ambitious program of distributing samples and specimens that would result in the Bruderheim meteorite becoming one of the most thoroughly studied and widely distributed meteorites in the world—even the Vatican Meteorite Collection has a sample.[29] Only two weeks after the meteorite's fall, Folinsbee wrote to Ward's Natural Science Establishment, Inc., a major supplier of geological and biological specimens in Rochester, New York: "The University has purchased all available fragments of the [Bruderheim] fall....We would like to classify and catalogue the material and then make it available to museums or interested persons throughout the world. Do you handle this type of material?"[30] Ward's replied with alacrity that they "would be very much interested in having the opportunity to handle the distribution of the meteorite."[31] A first shipment of 3850 g sent to Ward's sold out within days: "We have been very much surprised to find that we were able to place all of the Bruderheim meteorites you sent us right away, and I wonder if you could send us an equivalent shipment or even a larger one in the very near future. "[32]

Folinsbee was determined to extract maximum benefit from the Bruderheim fall for the University of Alberta and for Canadian meteoritical science. His policy was to make the museum specimens available on an exchange basis, "in order to build up our own collection....Lower priority will be given to institutions wishing to purchase museum quality fragments."[33] He arranged exchanges with the U.S. National Museum in Washington, DC; the American Museum of Natural History in New York; the Peabody Museum at Yale University; and numerous other institutions around the world. Folinsbee dealt more kindly with more youthful requesters; 10-year-old Richard Cawthorn of Deep River, Ontario, wrote and asked for a piece of Bruderheim to examine

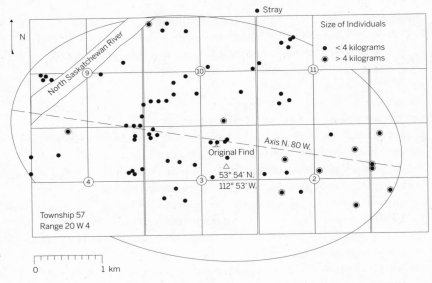

Fig. 8.6. The Bruderheim ellipse of fall. Note that the heavier (>4 kg) fragments are grouped at the southeastern apex of the ellipse.

with his microscope and enclosed 35 cents for return postage. Folinsbee sent him a small piece from specimen B-74 and a friendly letter. Three years later, after the Peace River meteorite fell, Colleen Dougherty of Edmonton wrote, saying: "I am 8 years old and I have been collecting rocks for a long time."[34] She asked for a "tiny bit" of the Peace River meteorite, but Folinsbee sent her two small fragments of Bruderheim instead. He explained in his reply: "The Peace River meteorite is very similar to Bruderheim, but so far only one piece has been recovered and it is required for scientific analysis."[35]

When Folinsbee sent a fragment of Bruderheim to the Redpath Museum at McGill University in Montreal, it was the first addition to that museum's meteorite collection since 1914.[36] Meteoritic research at the University of Alberta and in Canada as a whole received its initial impetus from the fall of the Bruderheim meteorite. The meteorite was efficiently recovered, distributed, and studied because of the presence in Edmonton of both astronomical and geological organizations, but it was realized that a similar event might have been poorly investigated or even have passed unnoticed in many parts of Canada. This realization led to the establishment on 31 May 1960 of the Associate Committee on Meteorites (ACOM) to act as a national clearinghouse for fireball and meteorite data, to co-ordinate and facilitate the study of Canadian meteorites, to promote public awareness of meteorites, and to advance the science of meteoritics in Canada generally. The formation of ACOM was spearheaded by the National Research Council in Ottawa, with representation from the University of Alberta and other Canadian universities, the Royal Astronomical Society of Canada, and several federal departments and agencies.

Mineralogy and Textures

Folinsbee was also busy sending samples from the Bruderheim specimen B-74 to some two dozen researchers in Canada, the United States, and Europe.[37] He would later suggest that they try to arrange for their papers to be published concurrently in the *Journal of Geophysical Research* with a paper being submitted by University of Alberta researchers.[38] The paper he was alluding to was a preliminary report on the chemistry and mineralogy of Bruderheim, by H. Baadsgaard, F.A. Campbell, and R.E. Folinsbee of the Department of Geology and G.L. Cumming of the Department of Physics. They identified the meteorite as a "typical gray chondrite," more specifically a *hypersthene* olivine chondrite (equivalent to a group L chondrite), with a somewhat higher alkali content than most chondrites and a low potassium-argon date of 1.60 Ga.[39] The analytical department at the Sherritt Gordon nickel refinery provided a check determination of the university's chemical analysis and another local company, Premier Steel, determined the carbon content (0.04%).[40] Campbell and Baadsgaard presented a paper on the Bruderheim meteorite at the 42nd Annual Meeting of the American Geophysical Union in Washington, DC, in April 1961. They gave slightly different chemical composition values; listed the silicate and metallic minerals in order of abundance; and reported potassium-argon dates for the whole meteorite (1.9 Ga), the feldspar fraction (1.6 Ga), and the pyroxene fraction (1.8 Ga).[41] Brian Mason (American Museum of Natural History) measured the fayalite fraction of the olivine in Bruderheim at 24%, confirming the meteorite's hypersthene-olivine chondrite classification.[42]

W.R. Van Schmus (University of Kansas) and P.H. Ribbe (Virginia Polytechnic Institute) used electron microprobe analysis and X-ray diffraction techniques to determine the chemical compositions and structural states of feldspar in some 30 chondrites. They found that the feldspar in Bruderheim consisted of a solid solution (mixture) comprising 83.5% albite (sodium-rich feldspar), 10.8% anorthite (calcium-rich), and 5.7% orthoclase (potassium-rich).[43]

One of the recipients of a Bruderheim sample was a research team at the California Institute of Technology in Pasadena led by Michael Duke. They deemed it "desirable to make an independent petrographic examination of the meteorite together with an independent chemical analysis. This work was purposely carried out in its entirety without knowledge of the results of other workers."[44] The Duke team prepared one polished section and one polished thin section for microscopic examination. They identified the nonopaque minerals of the meteorite as primarily olivine and pyroxene with minor amounts of feldspar, *apatite*, and *merrillite* (a calcium phosphate mineral, synonymous with *whitlockite*). The opaque minerals are metallic iron-nickel (Fe-Ni), troilite, and *chromite*. Well-developed chondrules and poorly developed, indistinct, or fragmental chondrules were found in equal numbers, and the former were examined and classified in detail (Fig. 8.7). Modal data were obtained by measuring nonopaque minerals in transmitted light and opaque minerals in reflected light. The Bruderheim meteorite contains 46% pyroxene and 43% olivine by volume (42% and 41% by weight, respectively). It contains 3.5% Ni-Fe metal and 5.1% troilite by volume (7.6% and 6.7% by weight, respectively). Chemical analyses were made by a variety of methods on crushed composite and separated (magnetic and non-magnetic fractions) samples.

The use of the term *opaque minerals* to describe the appearance of meteoritic metallic grains seen in a *transmitted* light microscope hints at the analytical limitations of this tool. However, such minerals can be examined very well in a *reflected* light microscope, and an experienced microscopist can identify hundreds of minerals; research microscopes commonly have both options. Initial measurements of the compositions of individual metal particles made in the mid-1960s, using electron microprobes that were in the early stages of development, were an improvement, giving some indications of the variability of the Ni/Fe ratio. However, these early methods of quantitative analysis were so time consuming as to deter extensive investigations. A quarter of a century later, modern instruments and techniques made it possible for Dorian G.W. Smith (University of Alberta) and S. Launspach (Sherritt Gordon Limited, Fort Saskatchewan, Alberta) to examine in great detail the compositions and textures of metal particles in Bruderheim.[45] They investigated intra-grain compositions of several metal particles in a polished thin section of Bruderheim, finding considerable variations in Ni, cobalt (Co), and Fe levels over short distances within individual grains. Nickel content ranged from 3.2% in kamacite to 58% in tetrataenite; Co ranged from 0.0% to 0.73% independently of Ni content; some taenite grains had Ni-enriched rims, a feature noted by earlier investigators. Controversy over the origin of metal particles in chondritic meteorites centres around whether they are of primary origin, that is, they were formed in the primordial solar nebula and acquired their characteristics prior to accretion, or whether they are are of secondary origin, that is, they were formed *in situ* after accretion, acquiring their characterisitcs by heating and

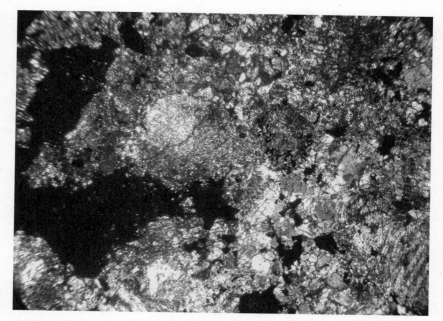

Fig. 8.7. A polished thin section of the Bruderheim meteorite seen in transmitted cross-polarized light. The large chondrule near the centre shows signs of shock; the adjacent smaller, darker chondrule is nearly intact. Note the opaque minerals (black). (31× magnification). Field of view 4.9 mm. For colour image see p. 288.

recrystalization. Smith and Launspach sought a model "which will allow the heterogeneities found among the silicate phases of L3 chondrites to be virtually eliminated by equilibration in types 6 (e.g., the Bruderheim meteorite) and 7, while leaving metal grains largely unaffected."[46] They envisaged a model in which the thermal event (heating) that homogenized individual silicate phases left the much less abundant metal phases unaltered because individual grains were too isolated to reach equilibrium with each other.

Many minerals exist in several forms, distinguished by the proportion of certain elements in their crystalline structure; for example, the proportion of calcium (Ca) in pyroxene ranges between high-Ca content clinopyroxene and low-Ca orthopyroxene. In both these cases, the elemental proportions stabilize, or *equilibrate*, at different concentrations depending upon temperature and pressure. By measuring Ca levels in the pyroxene of a meteorite, it is possible to estimate the temperature at which equilibration occurred in that meteorite. Edward J. Olsen (Field Museum) and T.E. Bunch (NASA Ames Research Center) used three different so-called *mineral-thermometers*, developed by other researchers, to estimate the equilibration temperatures of 24 H6, L6, and LL6 chondrites.[47] Olsen and Bunch calculated equilibration temperatures for Bruderheim (average of two samples) of 895 °C, 935 °C, and 1000 °C, all of which are slightly higher than the mean for all L6 chondrites in their study.

The violent collision between a parent body and another asteroid creates certain conspicuous shock-induced features in meteorites. A shock classification scheme for *equilibrated chondrites* (see Chapter 11) developed by R.T. Dodd (State University of New York, Stony Brook) and E. Jarosewich (U.S. National Museum) distinquishes six *facies* of shock (a, b, c, d, e, and f) of increasing severity; Bruderheim is assessed as facies d.[48] In their proposed shorthand terminology, it would classified as a L6d chondrite. A later and more widely adopted classification scheme (see Chapter 6) proposed by Dieter Stöffler and his colleagues at the University of Hawaii catalogues and organizes meteoritic shock features into a scale of progressive levels, or stages, of shock.[49] They classified Bruderheim as S4 (moderately shocked).

There is an ongoing effort by meteoriticists to identify the parent-objects of meteorites, but this is difficult; whereas meteorites can be analyzed in the laboratory to determine their exact chemical composition and petrologic type, most of the information about asteroids comes from astronomers using telescope-based spectroscopy. A team of researchers from Cornell University and the Grumman Aerospace Corporation in NewYork studied the light-scattering properties of the Bruderheim meteorite with the hope of matching these with the characterisitics of a known asteroid. W.G. Egan, J. Ververka, M. Noland, and T. Hilgeman crushed, ground, and sieved a piece of Bruderheim to prepare four samples for analysis: a coarse sample consisting of 0.25–4.76 mm fragments; a "coated" sample of the same 0.25–4.76 mm fragments covered with fine (<37μm) Bruderheim dust; a medium sample of 75–250 μm fragments; and a fine sample (<37 μm).[50] They used a goniometer that consisted of a light source on a 3 m long arm and a photometer on another 3 m long arm, both hinged so as to rotate about a central 60 mm diameter flat sample holder (Fig. 8.8), to measure the spectral reflectance, photometric, and polarimetric properties of the samples. Briefly, they found that the <37 μm sample reflected more light than the 0.25–4.76 mm sample, with the other samples intermediate; that is, when exposed to three different colours of light (ultraviolet, λ = 0.36 μm; green, λ = 0.54 μm; and red, λ = 0.67 μm), all four samples reflected less light (became darker) with increasing angle of incidence (see Fig. 8.8), and the amount of polarization increased as the wavelength of the light and angle of incidence increased. In attempting to identify the asteroid parent body of Bruderheim, the spectral reflectance and polarization data are more useful than the photometric measurements because asteroid photometric observations involve integrated or averaged light values for the whole disk of the asteroid, whereas laboratory measurements simulate conditions at only a single point on the disk. Comparing their Bruderheim data with asteroid data published by other researchers, Egan and his colleagues found that only Vesta and Icarus have spectral reflectance and polarization curves at all similar to Bruderheim. They were forced to consider the possibility that ordinary chondritic material may be rare as a constituent of the surfaces of large asteroids in the main asteroid belt but is very common in asteroids with Earth-crossing orbits.

Another team of Cornell researchers, Jonathan Gradie, Joseph Veverka, and Bonnie Buratti, made a more detailed examination of the problems inherent in matching laboratory measurements of the light-scattering properties of meteoritic material with astronomical measurements of asteroids.[51] They found that the normal reflectance of powdered Bruderheim material is 0.26 (26%) in green light ($\lambda = 0.55$ μm) for 45–74 μm particles.[52]

Bulk Chemistry

To a geologist or meteoriticist, *major* elements are the eight abundant rock-forming elements: oxygen, silicon (Si), aluminum (Al), iron (Fe), Ca, sodium (Na), potassium (K), and magnesium (Mg), although Bruderheim contains more Ni than either Na or K. *Minor* and *trace* elements are generally synonymous, although sometimes minor elements are considered to be those present in the range of 0.1%–1%, and trace elements are those with concentrations less than 0.1% or, occasionally, less than 0.01%.

James R. Vogt and William D. Ehmann (University of Kentucky), another research team to receive an early sample of Bruderheim from Folinsbee, determined the Si abundance in 107 chondrites and 11 achondrites by neutron activation analysis (NAA).[53] They measured the Si content of Bruderheim as 17.9% by weight, somewhat lower than earlier results by Baadsgaard et al.[54] and Duke et al. (Table 8.1).[55]

R.A. Schmitt, G.G. Goles, R.H. Smith, and T.W. Osborn (Gulf General Atomic and the University of California, San Diego) determined abundances of several elements in more than one hundred stony meteorites, including Bruderheim, by instrumental neutron activation analysis (INAA).[56] Although Schmitt and his team were able to test only single samples of most meteorites, in the case of Bruderheim five samples were available; they reported averaged abundances of Na, scandium (Sc), chromium (Cr), manganese (Mn), Fe, Co, and copper (Cu) in Bruderheim.

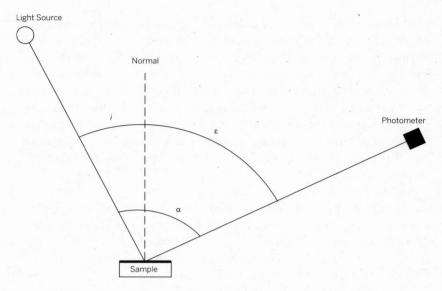

Fig. 8.8. Schematic diagram of a goniometer. The incidence angle, *i*, and the emission angle, ε, are varied to produce phase angles, α, ranging from 4° to 120°.

Masako Shima and Masatake Honda (University of Tokyo) determined the distribution of nearly two dozen alkali, alkaline earth and rare earth elements in crushed samples of three Alberta meteorites—Bruderheim, Peace River, and Abee—by NAA (Tables 8.1 and 8.2).[57]

Minor and Trace Elements

In 1973, another team of University of Tokyo chemists, Akimasa Masuda and Noboru Nakamura, with Tsuyoshi Tanaka from the Geological Survey of Japan made a new, more accurate determination of rare earth elements (REE) in 10 chondrites.[58] They used a mass spectrometric isotope dilution technique following chemical dissolution of the meteorite samples and separation of REE from major elements. Masuda and Nakamura measured the absolute abundances of the elements lanthanum (La), cerium (Ce), neodymium (Nd), samarium (Sm), europium (Eu), gadolinium (Gd), dysprosium (Dy), erbium (Er), and lutetium (Lu) in Bruderheim (Table 8.2b). They found that all of the chondrites differ in relative REE abundances, so they chose to normalize these to the Leedey (Oklahoma) meteorite. They found Bruderheim to be similar to Leedey after normalization. It has become the convention, more recently, to normalize REE abundances in ordinary chondrites to C1 carbonaceous chondrites.

Table 8.1. Major and minor elements in the Bruderheim meteorite.

Method	Si (mg)	Al (mg)	Fe (mg)	Ca (mg)	Na (mg)	Mg (mg)	Mn (mg)	Ni (mg)	Cr (mg)	P (mg)	K (μg)	Reference
NAA	179											Vogt and Ehmann 1965
Flame photometry, etc.	186	9.8	227	12.4	7.5	150			4.1		1080	Baadsgaard et al. 1961
WDCA	185	11.4	223	12.7	7.3	149		12.2	3.6	1.2	996	Duke et al. 1961
INAA			215			145		13.0				Sears and Weeks 1986
FP				8.6	7	148					910	Shima and Honda 1967b
INAA			231		6.7		2.4		3.8			Schmitt et al. 1972
NAA								12.7				Müller et al. 1971

Note: Concentrations are given in units per gram of meteorite. Method abbreviations are as follows: NAA, neutron activation analysis; WDCA, wet and dry chemical analyses; INAA, instrumental neutron activation analysis.

Zirconium (Zr) and hafnium (Hf) are also of great interest to geochemists and meteoriticists. These two elements have nearly identical chemical properties and nearly the same atomic and ionic radii, in spite of their dissimilar atomic weights (Zr, 91.224; Hf, 178.49). Measuring the Zr/Hf ratio in meteorites and terrestrial materials is important to the determination of a scale of relative cosmic abundances because the primordial Zr/Hf ratio is among those least likely to have been altered by chemical differentiation processes operating since the formation of the solar system. J.L. Setser and W.D. Ehmann (University of Kentucky) measured Zr and Hf levels in Bruderheim by NAA (Table 8.2a).[59] Setser and Ehmann reported a Zr/Hf ratio of 210 in Bruderheim compared with an average meteoritic Zr/Hf ratio of approximately 200 and an average terrestrial ratio of 39. They suggested that the Zr/Hf ratio "almost uniquely distinguishes between known meteoritic and terrestrial materials for the limited number of samples analyzed in this work."[60] Twelve years later, R. Ganapathy, Grace M. Papia, and Lawrence Grossman (University of Chicago)[61] largely repeated the work of Setser and Ehmann but made greater efforts to avoid contamination and sampling errors. Ganapathy et al. confirmed the earlier Hf results, but they reported much lower levels of Zr, with resultant Zr/Hf ratios near 32 (Bruderheim = 33.0) for both meteoritic and terrestrial materials.[62]

The concentration of the six platinum group metals (ruthenium, Ru; rhodium, Rh; palladium, Pd; osmium, Os; iridium, Ir; and platinum, Pt) in acid-dissolved samples of several stony and iron meteorites was measured spectrophotometrically by J.G. Sen Gupta of the Geological Survey of Canada.[63] He found the following levels in Bruderheim: 0.25 ppm Rh, 0.50 ppm Pd, 0.53 ppm

Table 8.2a. Trace elements in the Bruderheim meteorite.

Method	Co (μg)	Mo (μg)	Li (μg)	Rb (μg)	Sr (μg)	Ba (μg)	Zr (μg)	Cu (μg)	Zn (μg)	Ga (μg)	Ge (μg)	As (μg)
Flame photometry, etc.	500											
NAA												
INAA	442									5.7		1.1
INAA			1.6[a]	2.8	10.5	4.3						
INAA	640							95				
NAA							40					
RNAA							6.6					
NAA	5700[b]	6.38[b]										
NAA										5.4	10.1	
NAA								107	50			
Spectrophotometry									50			
Spectrophotometry											8.55	
NAA												
NAA												
Resonant nuclear reaction												
NAA												
NAA												

Note: Concentrations are given in units per gram of meteorite. Method abbreviations are as follows: NAA, neutron activation analysis; INAA, instrumental neutron activation analysis; RNAA, radiochemical neutron activation analysis.

[a]Lithium was determined by flame photometry.

[b]Metal phase only; for silicate and troilite phases, see paper.

[c]Mean of four samples.

[d]Mean of two samples.

Os, 0.40 ppm Ir, and 0.60 ppm Pt, but repeated efforts failed to detect any Ru (Table 8.2b). His data were included in Brian Mason's compilation *Handbook of Elemental Abundances in Meteorites* (1971). Mason and University of Virginia chemist Ralph Allen subsequently compared the relative abundances of the 14 stable rare earth elements La, Ce, praseodymium (Pr), Nd, Sm, Eu, Gd, terbium (Tb), Dy, holmium (Ho), Er, thulium (Tm), ytterbium (Yb), and Lu in the separated pyroxene, plagioclase, and phosphate mineral phases of Bruderheim and several other meteorites.[64]

In 1947, the German physicist Hans E. Suess proposed that the relative abundance of each element in the solar system depends in a fairly regular way on the element's mass and that this abundance pattern was caused by a

Table 8.2a. *(continued)*

Method	N (µg)	F (µg)	Cl (µg)	Br (µg)	I (ng)	Sn (ng)	Hf (ng)	Reference
Flame photometry, etc.								Baadsgaard et al. 1961
NAA								Ehmann et al. 1961
INAA								Sears and Weeks 1986
INAA								Shima and Honda 1967b
INAA								Schmitt et al. 1972
NAA							190	Setser and Ehmann 1964
RNAA							200	Ganapathy et al. 1976
NAA								Imamura and Honda 1976
NAA								Tandon and Wasson 1968
NAA								Greenland and Goles 1965
Spectrophotometry								Nishimura and Sandell 1964
Spectrophotometry						850		Shima1964
NAA		117[c]						Reed 1964
NAA			0.5–50	5–25	450			Reed and Allen 1966
Resonant nuclear reaction		32						Allen and Clark 1977
NAA	14							Kothari and Goel 1974
NAA			100[d]	1.25[d]				Wyttenbach et al. 1965

combination of nuclear properties and the way heavy elements are created in stars (nucleosynthesis). Harold Urey, who won the Nobel Prize in 1933 for his discovery of deuterium, later became interested in planetary science. In 1956, Suess and Urey, then at the University of Chicago, together published a seminal paper on the abundances of the elements based on meteorite data, combining Urey's knowledge of meteorite chemical data and Suess's ideas on how abundances should be interpreted.[65] However, at the time of the Bruderheim meteorite's fall, the atomic abundances of many of the heavy elements, including tungsten (W), were still uncertain. A team of University of Kentucky chemists led by W.D. Ehmann and A. Amiruddin, who were early recipients of Bruderheim material, made the first neutron activation analyses of tantalum

Table 8.2b Trace elements in the Bruderheim meteorite.

Method	Cs (ng)	La (ng)	Sm (ng)	Eu (ng)	Yb (ng)	Lu (ng)	Nd (ng)	Gd (ng)	Ce (ng)	Er (ng)	Tb (ng)	Dy (ng)
Mass spectrometry		378	247	84.1		39.1	765	33.5	1031	271		
INAA	<10	610	270	90	240	36	780	360			64	370
Spectrophotometry												
NAA												
NAA												
INAA												
NAA												
NAA												
NAA												
NAA												

Note: Concentrations are given in units per gram of meteorite. Method abbreviations are as follows: NAA, neutron activation analysis; INAA, instrumental neutron activation analysis.

[a]Order of magnitude determination.

(Ta), W, and Ir in this meteorite.[66] They reported their results as concentrations (ppm) and as atomic abundances (number of atoms per 10^6 atoms of Si). Uncertainty in the Ta values would later be addressed with a new NAA determination by Ehmann.[67] Amiruddin and Ehmann also measured the abundance of W in some 28 meteorites, tektites, deep-sea sediments, and numerous other geological specimens by means of NAA.[68] On the basis of their analyses of 17 chondrites, they calculated a cosmic abundance for W of $0.1/10^6$ Si; in Bruderheim, they found the W abundance to be 0.13 ppm by weight, equivalent to an atomic abundance of $0.10/10^6$ Si.[69]

In 1962, P.R. Rushbrook and W.D. Ehmann commented that "the cosmic abundance of the element iridium is of particular interest" and so it was, although not for the same reasons it would be some years later (see Chapter 18).[70] The small number of previous measurements of Ir in meteorites and concerns over possible sample contamination prompted Rushbrook and Ehmann to make a new, rigorously quality-controlled determination by NAA. Three years later another University of Kentucky team, P.A. Baedecker and Ehmann, made a similar NAA Ir determination, but this time included tektites and a number of terrestrial materials ranging from deep-sea sediments to granite rocks.[71] They reported that "iridium abundances in...terrestrial materials were found to be extremely low [≤ 0.01 ppm] and near the limit of detectability for this technique." In addition to Bruderheim's Ir content, Baedecker

Table 8.2b *(continued)*

Method	Pd (ng)	Os (ng)	Ir (ng)	Pt (ng)	Au (ng)	Rh (ng)	Ta (ng)	In (ng)	Reference
Mass spectrometry									Masuda et al. 1973
INAA									Shima and Honda 1967b
Spectrophotometry	500	530	400	600		250			Sen Gupta 1968
NAA			510				40		Ehmann et al. 1961
NAA							17		Ehmann 1965
INAA			506		146				Sears and Weeks 1986
NAA			510						Rushbrook and Ehmann 1962
NAA			470	900[a]	170				Baedecker and Ehmann 1965
NAA			490						Müller et al. 1971
NAA			530					63	Tandon and Wasson 1968

and Ehmann also measured its Au and Pt. Otto Müller, Philip A. Baedecker, and John T. Wasson (University of California, Los Angeles) later reported a mean bulk concentration of Ir and Ni in Bruderheim, determined simultaneously by NAA.[72]

Keiko Imamura and Masatake Honda (University of Tokyo) measured the distribution of W, molybdenum (Mo), and cobalt (Co) in the metal, silicate, and troilite phases of Bruderheim and several other chondrites by NAA.[73] As might be expected, the siderophilic element Co was heavily concentrated in the metal phase of Bruderheim (5700 ppm), compared with silicate (12 ppm) and troilite (54 ppm). Imamura and Honda found the concentrations of W in metal, silicate, and troilite to be 1.60 ppm, 0.044 ppm, and 0.084 ppm, respectively, whereas the corresponding values for Mo were 6.38, 0.31, and 3.5 ppm, respectively.

S.N. Tandon and John T. Wasson made replicate NAA determinations of gallium (Ga), germanium (Ge), indium (In), and Ir in Bruderheim and other stony meteorites.[74] They found the mean concentrations of these elements in Bruderheim to be close to those of Peace River, substantiating the already-noted similarity between the two chondrites. They also found that In concentrations are strongly correlated with petrologic type, with severely shocked L6 chondrites such as Bruderheim and Peace River having very low levels of In and weakly shocked L3 chondrites having the highest levels.

Chondrites are aggregates of material that accreted in the solar nebula 4.6 Ga ago; during this accretion several chemical and physical processes occurred which produced the many different classes of chondrites. Derek W.G. Sears and Karen S. Weeks (University of Arkansas) identified four such processes and investigated the process involved in the physical separation of metal and silicate, resulted in the siderophile element differences between the CV and CO classes of carbonaceous chondrites, the EH and EL classes of enstatite chondrites, and the H, L, and LL classes of ordinary chondrites.[75] They determined the abundances of Fe, Ni, Co, Ir, Ga, As, and Mg by NAA in 38 type 3 (least metamorphosed) ordinary chondrites and 15 equilibrated chondrites including Bruderheim.

L. Greenland and Gordon G. Goles (University of California, San Diego) determined the concentration in crushed samples of Bruderheim of Cu (107 ppm) and zinc (Zn) (50 ppm) using a neutron activation technique employing radiochemical separations. The levels of Cu and Zn are similar to those of the other ordinary (H and L) chondrites tested but lower than the levels found in carbonaceous and enstatite chondrites.[76] Masakichi Nishimura and Ernest B. Sandell (University of Minnesota, Minneapolis) determined Zn in a wider range of meteorites spectrometrically with dithizone after separation from other metals and constituents by ion-exchange in hydrochloric acid solution.[77] They determined an average level of 54 ppm Zn in ordinary chondrites (Bruderheim = 50 ppm, in very good agreement with Greenland and Goles), most of it in the silicates. They found an average of 730 ppm in Abee, likely because of the greater presence of sulphides in Abee (Zn is a "sulphur-loving" element) than ordinary chondrites.

As we have seen, Folinsbee's ambitious distribution scheme resulted in the rapid and widespread dissemination of Bruderheim fragments to researchers. Museums that acquired samples for research purposes often redistributed them to other researchers. Ward's Natural Science Establishment in Rochester, New York, provided still another source of Bruderheim samples by sale or exchange. G.G. Goles and E. Anders (University of California, La Jolla), for example, obtained a small bit of B-137, a 957 g specimen originally obtained by Ward's as part of an exchange with the University of Alberta.[78] Goles and Anders determined the iodine (I), tellurium (Te), and uranium (U) content of 12 chondrites, including Bruderheim, by NAA (Table 9.2c).[79] Their results ranged so widely that they commented that "faced with the observed variation...among the meteorites analysed, one should be very cautious in deriving mean abundances or cosmic abundances from these data." [80] C.L. Smith and his colleagues at the Western Australian Institute of Technology noted that almost all the analyses of Te in meteorites were by NAA but pointed out the advantages of solid source mass spectrometry using the stable isotope dilution technique, including excellent sensitivity, high accuracy, and precision.[81] Using the latter method, Smith and his colleagues measured the concentrations of Te in some 28 meteorites, including Bruderheim.

Although NAA is often the method of choice in studying the chemical composition of meteorites, it requires access to a nuclear reactor for the irradiation of samples. This was a problem for meteoriticists at the time that Bruderheim fell, because the number of research reactors then available was limited. Some researchers turned to a much more accessible analytical instrument: the ubiquitous Beckman DU spectrophotometer.[82] In 1964, Masako Shima, then at the University of California, San Diego, measured the concentration of Ge and tin (Sn) in Bruderheim, by spectrophotometry.[83]

Since the accidental discovery of meteorites in Antarctica in 1969 by Japanese scientists, search teams from Japan, the United States, and Europe have recovered more than 25,000 meteorite specimens. A research team from the Max-Planck-Institut für Chemie in Mainz, Germany, led by Friedrich Begemann, analyzed the noble gas (principally, helium, He; neon, Ne; and argon, Ar) content of 11 of these meteorites and several others, including a *Berkeley Standard* sample of Bruderheim provided by Dr. J.H. Reynolds (University of California, Berkeley) for comparsion.[84] The Begemann team's measurements of ^{3}He, ^{21}Ne, and ^{38}Ar in Bruderheim agreed to within 1% with previous results by the same group and to within 5% with most results reported by other laboratories, thus substantiating their Antarctic meteorite analyses. Ludolf Schultz and Hartwig Kruse, also of the Max-Planck-Institut für Chemie, subsequently compiled a list of all known reports of meteoritic He, Ne, and Ar concentrations, including those for Bruderheim and five other Alberta meteorites.[85]

The light elements lithium (Li), beryllium (Be), and boron (B) are of particular interest because they provide clues about the conditions in the presolar nebula from which our Sun and its planets formed. In 1962, Claude Sill and Conrad Willis at the U.S. Atomic Energy Commission in Idaho Falls reported the development of an improved fluorometric procedure for determining sub-microgram concentrations of Be and the results of their measurement of several meteorites and rocks.[86] They found the average Be concentration for 13 chondrites and 1 achondrite was 0.038 ppm or $0.64/10^6$ Si atoms; that of Bruderheim was 0.034 ppm, whereas terrestrial granite averaged about 2.75 ppm and a sample of shale averaged 6.14 ppm. Hisao Nagai (Nihon University, Tokyo) and his co-workers measured ^{10}Be and ^{26}Al in four iron meteorites and two chondrites, Bruderheim and Peace River; in these last two, metal fractions were separated from stone fractions, and both were analyzed.[87] Bruderheim was found to contain: in metal, ^{10}Be (4.96 ± 0.33 dpm/kg, or disintegrations per minute per kilogram of sample), ^{26}Al (6.24 ± 0.79 dpm/kg); in stone, ^{10}Be (25.4 ± 1.7 dpm/kg), ^{26}Al (79.8 ± 5.7 dpm/kg). Nagai et al. also determined the following isotopic ratios in metal: ^{10}Be/^{9}Be ($2.02 \times 10^{-11} \pm 0.14 \times 10^{-11}$), ^{26}Al/^{27}Al ($3.67 \times 10^{-11} \pm 0.46 \times 10^{-11}$), ^{26}Al/^{10}Be ($1.15 \pm 0.19$ dpm/dpm); in stone, ^{10}Be/^{9}Be ($28.2 \times 10^{-11} \pm 1.9 \times 10^{-11}$), ^{26}Al/^{27}Al ($16.7 \times 10^{-11} \pm 1.2 \times 10^{-11}$), ^{26}Al/^{10}Be ($3.13 \pm 0.31$ dpm/dpm). Mingzhe Zhai and Denis M. Shaw at McMaster University determined the average B concentration of the Bruderheim meteorite as 0.87 ppm.[88] Zhai and Shaw, in collaboration with Eizo Nakamura and Toshio Nakano (Okayama University), later measured the isotopic ratio ^{11}B/^{10}B in Bruderheim as 4.08359, among the highest measured.[89]

B.K. Kothari and P.S. Goel (Indian Institute of Technology, Kanpur) measured the total nitrogen content in several types of meteorites including chondrites by NAA. They found the bulk N content of Bruderheim to be 14 ppm (mean of four samples).[90]

Adjacent to the noble, or chemically inert, gases (He, Ne, Ar, Kr, Xe, and Rn) in the periodic table are the halogens. Halogens (fluorine, F; chlorine, Cl; bromine, Br; iodine, I; and astatine, At) are the most electronegative elements and, because of their high chemical reactivity, are of considerable interest to researchers studying the chemistry of meteorites. George W. Reed and Ralph O. Allen (Argonne National Laboratory and University of Chicago) measured the Cl, Br, and I contents in samples of several classes of chondrites by NAA; from this work, they were also able to derive the U and Te contents.[91] They found that Cl is very heterogeneously dispersed in the ordinary chondrites, probably concentrated in at least two phases, including *chlorapatite*; in Bruderheim total Cl content ranged from ~2.5 to 5.0 ppm, although one sample measured 50 ppm. In contrast to Cl, the Br contents of ordinary chondrites did not vary greatly; in Bruderheim, all total Br contents were within the ~0.5 to 2.25 ppm range. Although the I content of the ordinary chondrites tested was fairly constant at ~60 ppb, one Bruderheim result was an exceptional 450 ppb. Reed and Allen also determined the Te and U content of Bruderheim. A. Wyttenbach, H.R. von Gunten, and W. Scherle (Eidg. Institut für Reaktorforschung, Würenlingen, Switzerland) also measured the Br content in Bruderheim and a few other stony meteorites by NAA.[92] Their results are similar to those of Reed and Allen. However, their Cl results, as reported in a concurrent paper, are much higher than those of Reed and Allen.[93] George W. Reed also used a neutron activation technique to measure the F content in several stony meteorites.[94] According to Reed, the fluorine might be associated with *fluorapatite*, or possibly another phosphate mineral, *whitlockite* (a calcium phosphate mineral, synonymous with merrillite). Ralph O. Allen and Patrick J. Clark (University of Virginia) also measured F in several terrestrial rock samples and some 20 meteorites, including Bruderheim.[95] Instead of the more usual NAA, Allen and Clark used a proton beam technique in which fluorine is transmuted into oxygen and they determined the total F content of Bruderheim to be 32 ppm, lower than that measured by Reed.

Hideki Masuda, Masako Shima, and Masatake Honda (University of Tokyo) measured the distribution of U and thorium (Th) by NAA in powdered samples of four chondrites including Bruderheim.[96] In Bruderheim, U was found to be concentrated in more soluble minerals such as sulphides, phosphates and some silicates such as olivine, whereas Th was concentrated in more resistant minerals such as plagioclase and pyroxene. Some fissionable elements, U and Th among them, yield some fission products that decay by neutron emission following β decay. These delayed neutron emitters have half-lives in the range 1–60 seconds in contrast to the prompt neutrons emitted during during the fission process. George L. Cumming (University of Alberta) used a delayed neutron counting technique to determine U and Th in the Bruderheim, Peace River, and Stannern (Czech Republic) meteorites.[97] Ghislaine Crozaz at Washington University in St. Louis mapped the U and Th distribution in Bruderheim and several other stony meteorites by a technique in which a polished section of meteorite is covered with a sheet of mica and this assemblage is irradiated as in NAA.[98] The fission fragments produced by neutron bombardment leave tracks in the mica; etching the mica with acid renders these *fission tracks* visible under a microscope, thus tracing out the U distribution. Crozaz also measured Th levels with hitherto unattainable precision. In Bruderheim, she found, U is found enriched only in chlorapatite, which usually occurs in aggregates of small (~30 μm) grains; the Th/U ratio in this phase is 4.4 ± 0.6. The mineral whitlockite (=merrillite) is present but is not significantly enriched in U.

Stable Isotopes

Sulphur (S) has four stable isotopes; of these, ^{32}S and ^{34}S are by far the most abundant (95.02% and 4.21%, respectively). M. Shima and H.G. Thode (McMaster University, Hamilton, Ontario), citing earlier research, pointed out that the $^{32}S/^{34}S$ ratio (~22.2) in meteorites is remarkably constant, with a

Table 8.2c Trace elements in the Bruderheim meteorite.

Method	Th (ng)	U (ng)	Te (ng)	I (ng)	Cs (ng)	Be (ng)	B (ng)	W (ng)	Reference
NAA		24	500						Reed and Allen 1966
INAA					<10				Shima and Honda 1967b
NAA	34–40	10							Masuda et al. 1972
NAA								130	Ehmann et al. 1961
DNA	171	14.5							Cumming 1974
NAA								130	Amiruddin and Ehmann 1962
NAA		8–120	200–3400	5–570					Goles and Anders 1962
Fission track	2800–3400[a]								Crozaz 1979
NAA								1600[b]	Imamura and Honda 1976
Mass spectrometry			400						Smith et al. 1977
Fluorometry						34			Sill and Willis 1962
PGNAA							870		Zhai and Shaw 1994

Note: Concentrations are given in units per gram of meteorite. Method abbreviations are as follows: DNA, delayed neutron analysis; NAA, neutron activation analysis; INAA, instrumental neutron activation analysis.

[a]In chlorapatite phase.

[b]In the metal phase only; for silicate and troilite phases, see paper.

maximum difference of the order of 0.04% ± 0.1%; in contrast, the ratio varies by up to 10% in terrestrial samples.[99] They measured the $^{32}S/^{34}S$ ratios of the metallic, silicate, and mixed phases of Bruderheim by mass spectrometry, finding them to be in very close agreement with previous reports for other meteorites. Three years later, H.R. Krouse and R.E. Folinsbee (University of Alberta) made a similar determination of the $^{32}S/^{34}S$ ratio in the troilite phase of Bruderheim.[100] Because of the striking similarities between Bruderheim and the Peace River chondrite, they also tested the latter. The detected difference in the $^{32}S/^{34}S$ ratio ratio of the two meteorites was less than 0.01%, which was judged to be the limit of precision of the analyses.

Oxygen (O) has three stable isotopes, of which ^{16}O (99.76%) and ^{18}O (0.205%) are the most abundant. The $^{18}O/^{16}O$ ratio is currently of considerable interest, because it can be used as a proxy thermometer to measure past global climate changes on Earth. The heavier ^{18}O gets left behind when ocean water evaporates, particularly during periods of colder temperatures, so the $^{18}O/^{16}O$ ratio in air decreases. Therefore, the isotopic ratios of oxygen trapped in air bubbles in core samples of ancient glacier ice preserve a record of past temperatures (cold: ratio smaller; warm: ratio larger). The $^{18}O/^{16}O$ ratio also varies between classes of meteorites. In 1965, P. Taylor and his team at the California Institute of Technology in Pasadena made an oxygen isotope study of minerals in over 30 meteorites using mass spectrometry, with the expectation of learning more about the temperatures of formation of meteorites.[101] They found the following sequence of increasing ^{18}O content for coexisting minerals of stony meteorites: olivine, pyroxene, plagioclase, free silica (i.e., quartz). Except for the carbonaceous chondrites, the $^{18}O/^{16}O$ ratio of a given mineral is quite uniform in each meteorite class, but differences exist between classes of meteorites. Taylor et al. measured the ^{18}O enrichment, relative to a standard, in Bruderheim as olivine, $\delta = 4.5‰ \pm 0.3‰$; pyroxene, $\delta = 6.1‰ \pm 0.1‰$; plagioclase, $\delta = 6.5‰ \pm 0.0‰$; they calculated the whole rock enrichment as $\delta = 5.4‰$.[102]

Isotope Chronology

The radioactive isotope of carbon, ^{14}C, is perhaps the most familiar radioactive isotope because of its widespread use in dating archaeological relics. It can also be used to measure the terrestrial age of meteorites; when a meteorite falls to Earth, it is shielded from cosmic rays by the atmosphere and the cosmogenesis of ^{14}C is halted; the ratio of ^{14}C to ^{12}C begins to drop as the radioactive isotope decays (with a half-life of ~5700 years) to ^{14}N. In 1962, H.E. Suess and H. Wänke (University of California, San Diego) heated pulverized samples of 12 meteorites, including Bruderheim, to 1200 °C in a sealed oxygen atmosphere.[103] Thus, any carbon in the meteorites was converted to CO_2, which was extracted, purified, and tested for radioactivity using a counter. Suess and Wänke found the radiocarbon content of six recently fallen stone meteorites, including Bruderheim, to have a mean value of 48.2 dpm/kg (disintegrations per minute per kilogram); using this mean value Earth and 5580 years for the ^{14}C half-life, they calculated the time since six finds fell. They were surprised at the high terrestrial ages—one meteorite fell 14,000 ± 3000 years ago, another 20,000 or more years ago—as "it was generally assumed that stony meteorites would disintegrate through weathering and oxidation in an average time of a few years."[104] Bruderheim, with a known terrestrial age, had a radiocarbon content of 55.8 ± 3.0 dpm/kg. More recently (1984), Robert M. Brown and several colleagues at the Atomic Energy of Canada Limited laboratory in Chalk River, Ontario, together with Edward L. Fireman of the Smithsonian Astrophysical Observatory in Cambridge, Massachusetts, measured the ^{14}C content of Bruderheim and several other chondrites, using a different method. They measured the radiocarbon content of Bruderheim as 49.8 ± 1.8 dpm/kg.[105]

The time when a parent body of the chondrites was disrupted by collision with another asteroid can be estimated from the gas retention age of L chondrites. As the radioactive ^{40}K in the parent body decays to ^{40}Ar, this gas is trapped, or retained, in the rock of the asteroid until the rock is heated by a major collision, causing outgassing or loss of the argon. G. Turner at the University of Sheffield, England, and J.A. Miller and R.L. Grasty at Cambridge University measured the $^{40}Ar/^{39}Ar$ ratio in Bruderheim and determined its gas retention age as 495 ± 30 Ma, at which time it lost around 90% of its *radiogenic* ^{40}Ar (Table 8.3).[106] The L chondrites from this event that are falling today were delivered to Earth by later collisions, some 25 Ma ago in the case of Bruderheim as measured by its cosmic exposure age. Outgassing of meteorites during thermal metamorphism may also be influenced by their gas permeability with concomitant effect upon gas retention age estimates. To assess this problem, Takafumi Matsui (University of Tokyo) and Naoji Sugiura and N.S. Brar (University of Toronto) measured the gas permeability of 11 ordinary chondrites, including some heavily shocked ones.[107] They found a positive correlation between porosity and gas permeability, with both being lower in heavily shocked chondrites than in mildly shocked chondrites, with the moderately shocked Bruderheim having an intermediate degree of porosity (8%) and gas permeability, indicating that shock tends to reduce porosity.

Masako Shima and Masatake Honda sought to determine the solidification ages of the Bruderheim, Peace River, and Abee meteorites by rubidium-strontium (Rb-Sr) dating.[108] This age-dating method is based upon the decay of ^{87}Rb to ^{87}Sr. Rubidium has a stable isotope, ^{85}Rb, and an unstable isotope, ^{87}Rb (half-life = 47.5 Ga), which decays by beta emission to ^{87}Sr.[109] Strontium has three other isotopes, ^{84}Sr, ^{86}Sr, and ^{88}Sr, which are stable and are not produced by radioactive decay of ^{87}Rb. As ^{87}Sr is produced, the ratio of ^{87}Sr to each of the other strontium isotopes increases. Shima and Honda determined the abundance and isotopic composition of Rb and Sr in the separated phases of the meteorites. By plotting their data on a graph, they were able to derive a solidification age of 4.54 Ga for Bruderheim.

The half-life of ^{87}Rb is so long that other age-dating methods, for example, uranium-lead (U-Pb), may offer greater precision. The uranium isotope ^{235}U (half-life = 740 Ma) decays to the stable ^{207}Pb; ^{238}U (half-life = 4.46 Ga) decays to the stable ^{206}Pb. The U-Pb system has another feature distinguishing it from the Rb-Sr or Nd-Sm dating methods in that an independent "clock" can be obtained by measuring only the Pb isotope composition, yielding a ^{207}Pb-^{206}Pb age not dependent on U concentration measurements. When the $^{207}Pb/^{204}Pb$ ratios of several samples of a meteorite are plotted against their

$^{206}Pb/^{204}Pb$ ratios, the data points fall on a line representing a single age; this is the so-called lead-lead dating method. N.H. Gale, J.W. Arden, and M.C.B. Abranches at Oxford University calculated from both U-Pb and Pb-Pb data that the formation age of Bruderheim is 4.536 ± 0.006 Ga and that the meteorite was disturbed (possibly by a shock event) at about 500 Ma ago, consistent with its Ar-Ar age.[110]

Meteoriticists compare the isotopic composition of xenon (Xe) in meteorites with that of terrestrial (atmospheric) Xe by using one isotope (e.g., ^{130}Xe or ^{132}Xe) as a reference point for comparing the relative abundances of the other isotopes. Any differences in isotopic abundances in meteorites relative to atmospheric abundances are referred to as *anomalies*. W.B. Clarke and H.G. Thode (McMaster University) determined the isotopic composition of Xe in Bruderheim by heating chunks and crushed samples of whole meteorite in a vacuum and analyzing the gases driven off in a mass spectrometer.[111] They found a positive ^{129}Xe anomaly, or enrichment, of 0.88 ± 0.03 and a negative ^{132}Xe anomaly, or depletion, of -0.37 (± 0.02). Anomalies can also exist in the separated mineral phases of a meteorite relative to the whole meteorite. These results were confirmed soon afterwards by Oliver K. Manuel and Marvin W. Rowe (University of Arkansas) who, in addition, measured the abundance and isotopic composition of helium, neon, argon, and krypton (Kr) in Bruderheim.[112]

Craig Merrihue (University of California, Berkeley) measured the isotopic composition of Xe in troilite and chondrules in Bruderheim by mass spectrometry of gases given off during *in vacuo* heating of samples.[113] He found the troilite to be depleted in Xe relative to the whole meteorite, with a negligible ^{129}Xe anomaly. The chondrules, which were carefully separated from the matrix by gentle crushing, are also depleted in total Xe but enriched in ^{129}Xe by a factor of 2.9. The discovery of such different compositions in two components of the same meteorite suggested to Merrihue that ^{129}Xe is produced by the *in situ* decay of ^{129}I previously incorporated in varying amounts into the several phases of the meteorite. In a later paper, based on his PhD thesis and published posthumously, Merrihue expanded upon his earlier work on Xe in Bruderheim.[114] He observed that Bruderheim is ideal for studies on separated meteorite phases because the large amounts of material available make it possible to separate testable amounts of even rare minerals. He commented prophetically: "Also, owing to this abundance, it promises to become one of the most analyzed of all meteorites." In addition to the troilite and chondrules Merrihue had previously studied, he was now able to determine Xe anomalies in the feldspar, pyroxene, and olivine phases and several further chondrule samples. Relative

Table 8.3. Age estimates of the Bruderheim meteorite.

Method	Solidification age (Ga)	Cosmic-ray exposure age (Ma)	Gas retention age (Ma)	Reference
$^{40}Ar/^{39}Ar$			495±30	Turner et al. 1966
$^{36}Cl/^{36}Ar$; $^{22}Na/^{22}Ne$		30±4		Honda et al. 1961
^{53}Mn		25.1		Englert and Herr 1978
$^{150}Sm/^{149}Sm$		28		Hidaka et al. 2000
Rb–Sr	4.54±0.16			Shima and Honda 1967a
U–Pb	4.536±0.006			Gale et al. 1980
$^{3}He/^{3}H$		40±30		Fireman and DeFelice 1961
$^{38}Ar/^{39}Ar$		36±4		Fireman and DeFelice 1961

to the whole total meteorite, Xe is enriched in the feldspar by a factor of 2.3 and depleted in the pyroxene, olivine, and troilite by factors of 1.3, 2.7, and 9, respectively. The ^{129}Xe isotope is enriched in the feldspar and pyroxene by factors of 5.5 and 1.2, respectively, and depleted in the olivine by 7.4 and in the troilite by 71.[115]

The enrichment of radiogenic ^{129}Xe derived from now extinct ^{129}I in chondrules likely resulted from the formation of iodine-bearing, high-temperature minerals in the chondrules as the latter condensed from a hot solar nebula. The matrix material, which made up the bulk of Bruderheim and other chondrites, formed later at lower temperatures. However, E.C. Alexander and O.K. Manuel (University of Missouri, Rolla) noted that Merrihue's bulk sample of Bruderheim (IBC-1) contained twice as much ^{129}Xe as the chondrule IBC-21.[116] To see if the non-magnetic matrix material makes a significant contribution to the ^{129}Xe observed in the bulk chondrites, Alexander and Manuel measured the abundance and isotopic composition of xenon in separated matrix material and chondrules from Bruderheim and another chondrite, Bjurbole (Finland). They found that, in Bruderheim, radiogenic ^{129}Xe is concentrated in the chondrules; in Bjurbole, the concentration is less pronounced. They suggested that the ^{129}Xe enrichment in Bruderheim chondrules might be the result of outgassing from the matrix caused by the reheating of Bruderheim by an impact event some 500 Ma ago; Bjurbole apparently did not lose ^{129}Xe this way. Therefore, it appears that matrix material of ordinary chondrites contains an appreciable fraction of the radiogenic ^{129}Xe reported in bulk chondrites.

W.B. Clarke and H.G. Thode also made "much more difficult" measurements of the isotopic composition of Kr in Bruderheim by heating meteorite samples in a vacuum and analyzing the gases driven off in a mass spectrometer.[117]

Although Clarke and Thode examined a total of only five different meteorites, they showed conclusively that there are differences in isotopic content between Kr from meteorites and Kr from the atmosphere. Bruderheim showed both the highest Kr content or abundance of the meteorites tested and the greatest Kr isotopic anomalies, relative to atmospheric krypton particularly for ^{78}Kr. They suggested that these anomalies are due to nuclear processes; for example, spallation reactions on Rb and Sr or neutron capture by Br, that affected meteoritic and atmospheric Kr differently.

Craig Merrihue also studied meteoritic Kr, a subject "largely neglected because of experimental difficulties."[118] Among the main impediments to research on Kr in meteorites was the problem of contamination; the ratio of Kr to Xe is 12 times higher in the air than in meteorites, so air contamination is a greater problem in measurements of meteoritic Kr than in measurements of meteoritic Xe.

Primordial noble gases often correlate with other volatile elements such as I, Te, C, In, bismuth (Bi), and thallium (Tl) in chondrites; that is, as concentrations of the latter increase or decrease, so do levels of the former. However, as R. Ganapathy and Edward Anders (University of Chicago) pointed out, "the true extent of this correlation is sometimes hard to judge, because noble gas and trace element data are rarely obtained on the same sample."[119] Using mass spectrometry, Ganapathy and Anders measured He, Ne, Ar, and Xe in powdered aliquots of 11 H chondrites, plus Bruderheim, whose trace element contents had been previously determined by other workers. They calculated cosmic-ray exposure ages for the Bruderheim meteorite (21 Ma) from the noble gas data and, from the combined trace element and noble gas data: 2.1 Ga using K-Ar and 1.0 Ga using U-He.

The Bruderheim meteorite fell to Earth two and a half years after the launch of Sputnik l, at a time of great scientific and public interest in all things pertaining to outer space. Meteoriticists, biochemists, and other researchers were busy analyzing certain rare meteorites called carbonaceous chondrites for signs of extraterrestrial life. They did not discover any non-terrestrial organisms, but they did find many organic (carbon-containing) compounds including hydrocarbons, amino acids, alcohols, and pigments called porphyrins. G.W. Hodgson, then at NASA's Ames Research Center at Moffat Field, California, and B.L. Baker at the Alberta Research Council in Edmonton examined eight chondrites and found porphyrins in four of them (all carbonaceous chondrites); the remaining four, including Bruderheim (an ordinary chondrite), did not contain any porphyrins.[120] The amount of ^{13}C, another carbon isotope, relative to ^{12}C in meteoritic organic matter can provide information on the origin of meteorites, as can the ratio of deuterium (^{2}H, "heavy" hydrogen) to ordinary hydrogen. In 1983, Jongmann Yang and Samuel Epstein at the California Institute of Technology reported measuring δ^{13}C and δ^2H by mass spectrometry in some 20 meteorites.[121] Yang and Epstein found δ^{13}C for Bruderheim = −23.7‰, well within the range observed on Earth. They concluded that, "If organic matter originated in the interstellar medium, our data would indicate that the ^{13}C/^{12}C ratio of interstellar carbon five [Ga] ago was similar to the present terrestrial value." On the other hand they found deuterium enrichments (δ^2H; Bruderheim = 93‰) as high as ~10 times terrestrial values, which they attributed to isotope exchange reactions (the replacement of hydrogen atoms in molecules by the heavier ^{2}H atoms) occurring at temperatures below 150 K (−123.16 °C).[122]

Tritium (^{3}H) is the third isotope of hydrogen; it is radioactive, with a half-life of 12.3 years. Cosmic-ray-produced ^{3}H in meteorites was first measured in 1957; within a few years, it was apparent that the ^{3}H content was much higher in stony meteorites than in most iron meteorites. Edward Fireman and James DeFelice at the Smithsonian Astrophysical Observatory (Cambridge, Massachusetts) suggested that stony meteorites are less shielded from the energetic cosmic rays that produce radioactive isotopes than metallic meteorites.[123] Fireman and DeFelice were among the first researchers to receive samples of Bruderheim from Folinsbee. They measured the radioactivity of one piece only 52 days after the date of fall and were able to "count" its short-lived ^{37}Ar (half-life = 35.0 days); the counting continued for several weeks until only the ^{39}Ar (half-life = 268 years) and ^{3}H remained. Fireman and DeFelice calculated a ^{3}He-^{3}H cosmic-ray exposure age of 40 ± 10 Ma and a ^{38}Ar-^{39}Ar cosmic-ray exposure age of 36 ± 14 Ma for Bruderheim using data on the ^{3}H and ^{38}Ar content of Bruderheim provided by another researcher; these are in good agreement with ages calculated by others.[124] Fireman and DeFelice also stated that there are two explanations for the short exposure ages of meteorites: recent large-body breakup and space erosion; they favoured the latter. They estimated that an exposure age of 36 Ma corresponds to an erosion rate of approximately 1.0 cm/Ma for the Bruderheim meteorite. Somewhat later, A.E. Bainbridge, H.E. Suess, and H. Wänke at the University of California, San Diego, measured the cosmogenic ^{3}H content of Bruderheim at 495 ± 20 dpm/kg.[125] St. Charalambus and K. Goebel (Conseil Européen pour la Recherche Nucléaire [CERN], Geneva) measured the ^{3}H and ^{39}Ar content of Bruderheim.[126] They heated meteorite samples in a vacuum; separated out first the hydrogen (with its tritium component), then the Ar; and pumped these gases into Geiger-Müeller and proportional counters.[127] Tritium and ^{39}Ar , which is also radioactive, both decay (into

^{3}He and ^{39}K, respectively) by emitting β-particles which are "counted." From their measurements, Charalambus and Goebel were able to calculate a ^{3}H-^{3}He cosmic-ray exposure age of 25 (± 10%) Ma and a ^{38}Ar-^{39}Ar cosmic-ray exposure age of 31 (± 10%) Ma for Bruderheim, in very good agreement with the ages found by Fireman and DeFelice (above) and Honda et al. (below).

Meteoriticists often make a cast or model of a meteorite, sometimes even painting and weighting it to match the original, before it is broken up for distribution or analysis (see Chapter 14). However, there are other reasons for making meteorite replicas. M.W. Rowe and M.A. Van Dilla at the Los Alamos Scientific Laboratory received an early, 2 kg sample of Bruderheim from Folinsbee to investigate its gamma-ray spectrum.[128] They were developing a nondestructive analytical method that gives whole-sample results, minimizing sampling errors. They created four hard-shelled copies of the meteorite and filled them with different mixtures of powdered iron and various chemicals, including a measured and calibrated amount of radioactive substance, to create a potassium mock-up, a ^{22}Na mock-up, a ^{54}Mn mock-up, and an ^{26}Al mock-up. The gamma-ray spectra of these mock-ups were measured in a very sensitive gamma-ray scintilllating spectrometer. These spectra were then compared with the meteorite's spectrum. Rowe and Van Dilla's measured levels of ^{22}Na, ^{54}Mn and ^{26}Al in Bruderheim agreed quite well with those reported elsewhere, but their value for the potassium content was lower than that reported by Baadsgaard et al.[129]

Rowe, Van Dilla,, and E.C. Anderson continued to develop their gamma-ray spectrometry method and later used it to analyze a suite of 37 chondrites and achondrites, permitting the comparison of Bruderheim with other meteorites; of particular interest to us is Harleton (Texas), a L6 chondrite like Bruderheim, which was analyzed just 21 days after it fell. They found that Bruderheim contained almost twice as much ^{54}Mn, one and a half times as much ^{22}Na, and nearly one-third more ^{26}Al than Harleton, despite the extreme freshness of the latter.[130]

In contrast to the nondestructive analytical technique employed by Rowe and his colleagues, the team of M. Honda, S. Umemoto, and J.R. Arnold (University of California, San Diego) dissolved 850 g of powdered Bruderheim in hydrofluoric and sulphuric acid; three other meteorites were similarly treated.[131] The Bruderheim specimen was provided by Folinsbee, and the other meteorites were obtained from the aforementioned Ward's Natural Science Establishment. Honda and his co-workers then tested the resultant samples for the presence of 18 radioactive nuclides, namely: ^{7}Be, ^{10}Be, ^{22}Na, ^{26}Al, ^{36}Cl, ^{46}Sc, ^{44}Ti, ^{48}V, ^{49}V, ^{51}Cr, ^{53}Mn, ^{54}Mn, ^{55}Fe, $^{56+58}$Co, ^{57}Co, ^{60}Co, and ^{59}Ni. Levels of radionuclides were measured by counting of β, γ, and X-ray emissions. The data led Honda et al to conclude that the cosmic ray intensity, averaged over the half-life of each radionuclide, has been nearly constant. They calculated a cosmic-ray exposure age for Bruderheim of 30 Ma.[132]

The production of cosmogenic isotopes is influenced by several factors including a meteorite's chemical composition and its hardness or shielding from radiation, and temporal variations in cosmic ray levels. In the course of trying to better estimate the constancy of the cosmic ray flux, Otto Müller and his colleagues W. Hampel, T. Kirsten, and G.F. Herzog at the Max-Planck-Institut für Kernphysik in Heidelberg, Germany, made improved light noble gas, ^{26}Al and ^{53}Mn analyses of several chondrites.[133] In three Bruderheim samples, they measured average levels of ^{3}He, ^{4}He, ^{20}Ne, ^{21}Ne, ^{22}Ne, ^{36}Ar, ^{38}Ar, ^{40}Ar, ^{53}Mn, and Fe. Müller et al. estimated that cosmic-ray intensity had varied by less than 20% in the last three or four half-lives of ^{53}Mn (= 36–48 Ma), with a possible recent ($\leq$2.5 Ma) increase.

P. Englert and W. Herr, in their study of ^{53}Mn-derived exposure ages of chondrites, cite the exposure ages of several meteorites, including Bruderheim (25.1 Ma), studied by other researchers.[134] This is in fair agreement with an exposure age of 28 Ma obtained by a Japanese research team led by Hiroshi Hidaka from a study of samarium isotope shifts from ^{149}Sm to ^{150}Sm.[135]

At the University of New Brunswick in Saint John, Ian Cameron and Zafer Top measured the level of cosmogenic ^{26}Al in a 277 g Bruderheim individual from the Geological Survey of Canada's National Meteorite Collection (specimen no. 0220106), using a gamma-ray spectrometer.[136] They measured an activity of 58 $\pm$ 2 dpm/kg from which, using the method of Lavrukhina and Ustinova,[137] they were able to postulate an orbit for the Bruderheim meteorite with an aphelion in the range of 2.05–2.85 astronomical units (the distance from the Earth to the Sun); in other words, the orbit, at its farthest from the Sun, grazes the inner edge of the main asteroid belt.[138] G.K. Ustinova, V.A. Alekseev, and A.K. Lavrukhina examined several methods for determining the pre-atmospheric size of a meteorite on the basis of data on its cosmogenic radionuclides.[139]

Julian P. Shedlovsky, Philip J. Cressy, and Truman P. Kohman at Pittsburgh's Carnegie Institute of Technology compared the abundances and ratios of 16 cosmogenic radionuclides in four recently fallen ordinary chondrites, confirming the previously noted similarities between Bruderheim and Peace River—with the exception of ^{46}Sc, which was six times higher in the latter.[140] Cressy, later at NASA's Goddard Space Flight Center, provided a similar comparison with three more recently fallen stony meteorites.[141] In a subsequent paper, Cressy commented: "in order to use the [level] of ^{26}Al in stone meteorites for the estimation of shielding effects, terrestrial ages, short cosmic-ray exposure ages, and effects caused by variations in a meteorite's radiation environment, the effect of chemical composition on the production of ^{26}Al must be known."[142] He chemically analyzed powdered Bruderheim samples and measured them for ^{26}Al activity in a gamma-ray coincidence counter. The production rates of ^{26}Al from the principal target elements, Mg, Al, Si, S, Ca, and Ni+Fe, were calculated. Similarly, to calculate cosmic-ray exposure ages of meteorites from their noble gas concentrations, the effect of meteorite chemical composition upon the spallation production rate must be known for each noble gas isotope considered. D.D. Bogard and Cressy measured the concentrations of the cosmogenic noble gases ^{3}He, ^{21}Ne, and ^{38}Ar in eight different mineral phases of the Bruderheim chondrite; from these data, they were able to derive a table of ^{21}Ne and ^{38}Ar production rates for the main chondrite and achondrite classes.[143]

Summary

Bruderheim, comprised of many variously sized stones aggregating to over 300 kg, was ideally suited for the rapid and widespread dissemination of samples undertaken by Bob Folinsbee, while the freshness of the fall guaranteed its eager reception by researchers. As a consequence, this otherwise very ordinary chondrite has achieved great significance in meteoritics and cosmochemistry. It is used as something of a reference material in radiocarbon[144] and rare-gas research; for the latter, the Bruderheim "Berkeley Standard" is used to cross-calibrate analytical procedures between different laboratories.[145]

9

The Ferintosh Meteorite

History

The 1960s were the golden years for meteorite discoveries in Alberta—no fewer than seven were found in an eight-year span, a record unequalled in Canada until the twenty-first century. One meteorite was found in October 1965 on the farm of I.S. (Sam) Enarson, about 3 km northwest of the village of Ferintosh, during threshing operations (Fig. 9.1). Mr. Enarson picked it up near his granary about 200 m from the farmhouse.[1] It appeared to be an unusual type of rock and was brought into the Department of Geology at the University of Alberta by Mr. Enarson's nephew, D.A. (Donald) Enarson who was a student at the university.[2] On November 10, Professor Folinsbee interviewed Don Enarson who said that he thought his uncle would be happy to sell the meteorite for $100. He also believed that another piece of the same meteorite had been found several years earlier on his father's farm, one mile (1.6 km) north.[3] The next day, Folinsbee and Dr. Charles Stelk drove down to Ferintosh to interview Mr. Enarson and to arrange for purchasing the meteorite. They saw Mrs. Enarson first and she signed the bill of sale for $100. Later, Mr. Enarson showed them the spot where the meteorite had been recovered. Folinsbee and Stelck searched for a while for other pieces but did not find anything.[4] The other meteorite that was found on Dale Enarson's father's farm had subsequently been lost and has never been recovered.[5]

Science

The Ferintosh meteorite weighed 2201 g when it was found; within a week, Folinsbee had sent a report on its type and general appearance with a brief account of its discovery to Professor E.L. Krinov (Academy of Sciences of the USSR, Moscow), editor of the *Meteoritical Bulletin*.[6] The meteorite is a typical chondrite with well-developed regmaglypts (Fig. 9. 2). It is slightly iron stained on the outer surfaces from weathering and is a dark brown colour, rather than the jet black characteristic of a pristine stony meteorite (Fig. 9.3). The fusion crust is still intact. The sharply angular surfaces of the specimen suggest that it may be part of an individual that broke up during atmospheric flight; that is

Fig. 9.1. Mr. Ingemar Samuel Enarson holding the meteorite finder's certificate presented to him by the National Research Council of Canada's Associate Committee on Meteorites (ACOM), for his discovery of the Ferintosh meteorite.

consistant with D.A. Enarson's report of another specimen. It is speculated that the Ferintosh meteorite may have come from a bright bolide reported in this area in the early 1930s.

Apart from this initial examination, the Ferintosh meteorite remained uninvestigated for another 30 years. In 1996, D.G.W. (Dorian) Smith (University of Alberta) prepared polished thin sections for detailed optical and electron microprobe analyses. Using optical microscopy, he described the meteorite as a thoroughly recrystallized chondrite with the textural characteristics of petrologic type 6 (Fig. 9.4). Chondrules, although generally still clearly visible in outline, are no longer glassy. Both barred olivine and radiating orthopyroxene chondrules are common, and the latter, particularly, tend to be rust stained from the weathering of the meteorite. The intervening matrix material between the chondrules is well crystallized and contains irregular patches of clear and glassy *maskelynite*, plagioclase feldspar that has been transformed into natural glass by shock.

Smith found further evidence of shock in the presence of a shock fusion vein running diagonally through the polished section. The vein, similar in texture to the one found in the Peace River meteorite (see Chapter 11), contains glass, metal, and troilite. The sharp boundaries of the vein—mineral grains a

Fig. 9.2. The Ferintosh meteorite showing the well-developed "thumbprints," or regmaglypts.

Fig. 9.3. Another view of the Ferintosh meteorite. Note the sharp edges.

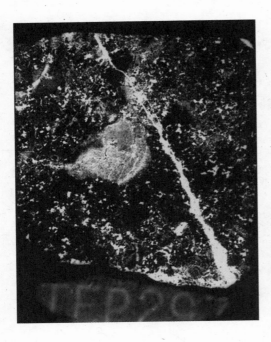

Fig. 9.4. The Ferintosh meteorite. The photograph was taken in unpolarized light by treating the polished sample (which is mounted on a glass slide) as a negative in a photographic projector. Opaque materials like metals, oxides, sulphides and the diagonal shock fusion vein appear white; clear, transparent minerals appear black. Other minerals, mainly silicates, appear as shades of gray.

few tens of micrometres from the vein show no apparent effects—indicate that the phenomenon, probably an impact with another meteoroid, was spatially very limited in its effects. Smith described the meteorite as weakly to moderately shocked.[7]

In his electron microprobe analysis of the chemical and mineralogical composition of Ferintosh, Smith acknowledged the limitations of energy dispersive analysis but showed how these could be compensated for. His results for olivine revealed a homogeneity consistent with the texture seen optically, with a mean fayalite content of 25.7%. The orthopyroxene showed a mean *ferrosilite* content of 21.2%, within the range normally observed in L-group chondrites. Only a few very small grains of clinopyroxene were located, and these had a mean content for *wollastonite*, enstatite, and ferrosilite of 43.2%, 46.3%, and 10.5%, respectively. The maskelynite analyses showed that a substantial amount of sodium (Na) had been lost during and after the shock-induced conversion of the original plagioclase feldspar. Merrillite (=whitlockite) commonly coexists with chlorapatite and can easily be distinguished from the latter in an energy dispersive spectrum by the absence of chlorine. In chondritic meteorites, some of the chlorine is replaced by fluorine. In Ferintosh, the fluorine content was not analyzed but was calculated stoichiometrically. Smith comments:

"Although no great store can be placed on the F contents derived in such a manner, it is interesting to see that the average value for the Ferintosh apatites is very close to those that are obtained by fully quantitative wavelength dispersive techniques for other L6 chondrites."[8]

Chromites fell within the ranges measured by others for L5 and L6 chondrites. *Ilmenites* and troilite are also similar to those reported elsewhere.[9]

Smith carried out a total of 646 point analyses for Fe, Ni, and Co on the metal grains in Ferintosh; Fe was detected by energy dispersive analysis while Ni and Co were detected by wavelength dispersive techniques, with all determinations being simultaneous and on exactly the same point. About 55% of the grains were found to be kamacite, the rest being taenite and plessite; this preponderance of kamacite is typical of the L group of chondrites.[10] Most of the kamacite grains contained about 7.5% Ni, the plessite grains about 16%, and the taenite grains mostly either 29% or 38% Ni. However, the Co content varied widely for a given Ni value; for example, the Co content of the kamacite ranged from ~0.3% to ~2.25%.

10

The Innisfree Meteorite

History

Flying above the clouds near Swift Current, Saskatchewan, on 5 February 1977, the crew of Air Canada's flight 167 from Winnipeg to Vancouver saw a brilliant fireball light up the night sky to the north. They were about 470 km from the meteor and radioed their sighting to the control tower in Regina, which promptly relayed it to Mr. John Hodges of that city. Mr. Hodges was a member of the National Research Council of Canada's (NRC) Associate Committee on Meteorites. He telephoned the Saskatoon office of the Research Council's Meteorite Observation and Recovery Project (MORP), which operated a network of sky cameras in western Canada (see Chapter 16). The MORP staff immediately called in the films from the camera stations at Watson, Asquith, and Neilburg in Saskatchewan and Vegreville in Alberta. It was generally cloudy in Saskatchewan, and none of the camera stations there photographed the meteor.

In Alberta, the sky was clear, and reports from fireball observers near Edmonton were directed to Professor R.E. Folinsbee at the University of Alberta. An appeal over local radio stations for observations produced over one hundred responses. The Fort Saskatchewan Detachment of the Royal Canadian Mounted Police (RCMP) also interviewed witnesses and sent their reports to Folinsbee. These reports came from distances of between 2 and 304 km from the point on the ground where the meteor was directly overhead at the time of maximum light. Six observers at four different locations were within 30 km of the groundpath, and another six were between 50 and 100 km distant, noted Ian Halliday, Alan Blackwell, and Arthur Griffin in their initial report.[1]

One of the more lengthy reports resulted from an interview by the RCMP with Miss Brendalee Walker, then age 13, who lived on a farm near Chipman, Alberta, about 77 km from the ground point beneath maximum meteor light. The report was made within an hour or two of the observation:

It was approximately 7:15 p.m. [5 February 1977], and I was standing outside of the farmhouse, near the woodbox. Our farm is 1–1/2 miles [2.4 km] north and 2 miles [3.2 km] east of Chipman, Alberta. I was looking north and I saw what appeared to be car lights shining off our gas tank, and when I looked up the whole east sky was lit up. There was a large spark and about six little ones trailing. They were heading east. They were all pure white and very bright. The large one was about the size of a large car and the rest were only small, about the size of basketballs. They were all in a straight line, going east and appeared to be only about 50 feet [15 m] over the top of the trees. I don't have any idea how fast they were going. I saw them for at least 30 seconds and they were still moving when I got to the house. The sparks appeared to be falling pretty rapidly and were something like a plane landing. The sparks appeared to be travelling about 40 miles per hour [64 km/h].

The police report accompanying Miss Walker's statement quotes her as also saying a swishing sound accompanied the lights, but there was no large sonic boom. However, she had run into her home before the sonic boom would have had time to reach her. Analysis of the photographic record reveals that she greatly overestimated the duration of the fireball, but this is a common tendency.[2]

Two other youthful witnesses, Lyell and Martin Ferguson, about ten and eight years old, were very close (possibly within 2 km) to the actual meteor path. They were walking south on a road about 14 km north-northeast of Innisfree when the sky lit up behind them. One of them turned around to look for an approaching car but the fireball was so nearly overhead he did not see it until he was again facing south, when he observed fragments descending high in the sky in front of them. Both boys described a sonic boom "like distant shooting dying away to the southeast about 1–1/2 minutes after the fireball."

The only other convincing report of sounds was from Mr. Ken McGilivray who was located somewhat northeast of Innisfree, about 11 km from the meteor path. He saw two large fragments plus six to eight small ones, and after going in the farmhouse to describe the meteor to family members, he again went outside. By later repeating his actions, he estimated that there was an interval of 1 minute 45 seconds between the meteor and the sound, which he described as "a rumble, then a sharp rumble, mellowed rumbling for 15 seconds, loud, then 5 seconds dying away." Another witness, in a schoolyard only 1 km further from the path, heard no sounds.[3]

Some witnesses were luckily located near the end and on either side of the fireball's path and analysis of the visual observations alone would have been sufficient to lead searchers to the right (albeit large) area. Clouds had obscured the sky over Saskatchewan, and none of the MORP camera stations

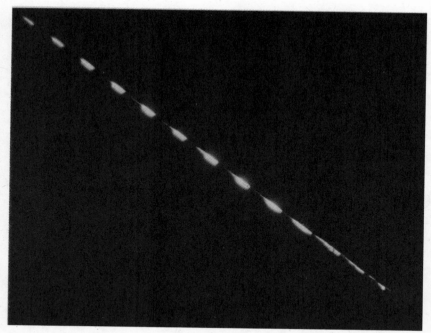

Fig. 10.1. The Innisfree fireball in flight, from upper left (at an altitude of 59 km) to lower right, photographed by the MORP camera at Vegreville, Alberta. A rotating filter wheel has chopped the meteor trail into four segments per second. Note the evidence of separate fragments near the lower end of the path.

there photographed the meteor. However the camera at Vegreville, Alberta, captured much of the fireball's flight on film (Fig. 10.1). Since at least two photographs, from two separate stations, were needed to calculate the meteor's trajectory, Eldon Hubbs, from the MORP office in Saskatoon, made the long drive to Lousana, Alberta, and back on the same day to retrieve that station's film. When the Lousana film was processed, a faint meteor image was found on the first exposure of the evening; it was of poor quality due to the bright twilight sky, but it was better than nothing.

The two meteor photographs were measured in Saskatoon, and the data were relayed by telephone to the NRC in Ottawa. By February 14, the NRC's computer had calculated an end point for the last luminous fireball fragment near a height of 20 km and a terminal velocity well below 4 km/s. The survival of this object to such a low altitude and velocity provided the first evidence that a meteorite fall was probable and made an immediate search desirable. Before starting a ground search, it was necessary to narrow the probable fall area. The computer was used to calculate the dark-flight (non-luminous) portion of the trajectory below 20 km, using atmospheric wind measurements made at Edmonton three hours before the meteor, values for the Earth's gravity and rotation, plus the drag or air resistance acting on a falling object at various heights. Fall locations were predicted for three hypothetical meteorites of average stony meteorite density and masses of 10, 4 ("P" in Fig, 10.3) and 0.5 kg. The computer predicted a spread between these three hypothetical fragments of only 2 km on the ground (the predicted area of fall based on only visual observations was a circle 20 km in diameter).

On February 16, Ian Halliday and Arthur Griffin flew from Ottawa to Edmonton and proceeded by car to Vegreville where they were joined by Alan Blackwell and Eldon Hubbs who drove from Saskatoon. That evening arrangements were made with young snowmobilers from Vegreville to begin a search next morning. The search began at 10:30 a.m. on February 17 in farming country about 13 km northeast of Innisfree, using four snowmobiles. three of them driven by youths with Halliday, Blackwell, and Griffin as passengers while the fourth was driven by Hubbs (Fig. 10.2). The snowmobiles with their drivers and lookouts traversed the fields in runs about 12 to 18 m apart, making frequent stops to examine field stones, clumps of black soil and animal burrows. At 4 p.m., Halliday spotted a dark object some 5 or 6 m to the side that, upon closer inspection, proved to be a 2.0 kg meteorite. The meteorite had evidently penetrated the 30 cm of snow, hit the hard frozen ground and rebounded to the surface, bringing up some soil and knocking a few small chips off the meteorite. Next day, an agreement was signed with Mr. William Fedechko, owner of the land on which the meteorite was found, for the purchase of the meteorite "for a fair price to be determined after preliminary study of the meteorite."[4]

After the first 2.07 kg fragment, no more meteorites were found despite a search covering nearly six square kilometres. Careful study of the Vegreville photograph during March 1977 strongly suggested that additional fragments had fallen, and the search effort was renewed in April once the snow had melted. On April 9, a small (33 g) broken-up piece was found by a group of students led by Folinsbee. It was located 600 m from the original find, on land owned by a neighbour of Mr. Fedechko.

Three days later, Mr. Fedechko notified the MORP office in Saskatoon that three pieces with a combined mass in excess of 1 kg had been found by his family and some visitors, all on the same quarter section as the original find. It appears that two of these were were found on April 10 and the largest of the three on April 11. The MORP staff from Saskatoon returned to Innisfree the

Fig. 10.2. Six of the seven members of the Innisfree meteorite search team. From left to right: E. Hubbs, V. Kuzz, A.T. Blackwell, K. Fried, I. Halliday and M. Freed. The meteorite can be seen in front of the third snowmobile from left.

following week, and a sixth piece of the meteorite was found by Blackwell on April 21. It was in numerous fragments, apparently the result of having been run over by a farm implement used in plowing a firebreak. The first piece found by the Fedechkos was in two fragments and the smaller of these was distributed as souvenirs, so that the total weight of this piece, estimated as 120 g, is uncertain. Pieces 4, 5 and 6, in order of discovery, weighed 345, 894, and 330 g. The total recovered mass was [at that time] 3.79 kg, and the mean co-ordinates of the fall were 111°20′15″W, 53°24′54″N. [5]

Later on, three more fragments (nos. 7, 8 and 9), weighing 22, 387, and 375 (?) g, respectively, were found by Mr. Fedechko; this additional material was purchased in 1978. The locations of all nine Innisfree fragments, having a combined mass of 4.576 kg, were mapped by Halliday, Griffin, and Blackwell (Fig. 10.3). [6]

Science

Fall Phenomena and Astronomical Interpretation

The *ellipse of fall* is unusully small, an area 500 × 1000 m can enclose the location of all the finds. The steep angle of the path through the atmosphere (67.8°) is the main reason the fragments dispersed over such a limited area.

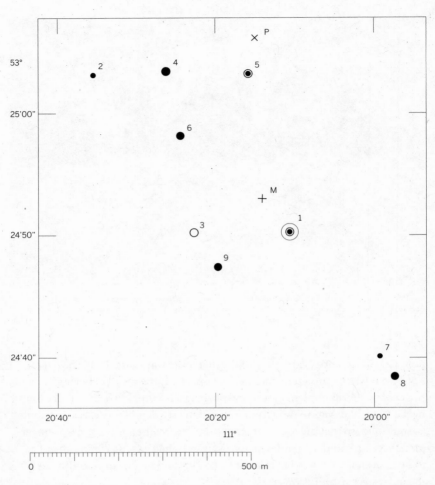

Fig. 10.3. Location map of the nine fragments of the Innisfree fall. The fragments are numbered in the order of their recovery. M is the mean position of the fall, and P is the predicted location for a hypothetical 4 kg fragment (see text).

The early portion of the meteor trail is missing from the Vegreville photograph (Fig. 10.1), because the MORP cameras did not cover the region near the zenith. The trail enters the camera field at a height of 58.8 km above sea level. Halliday and his colleagues estimated that the Vegreville trail would have been recorded 1.0 second earlier if it had been in the camera field, corresponding to a beginning height of 72 km for the Innisfree fireball, which had an entry velocity of 14.54 km/s.

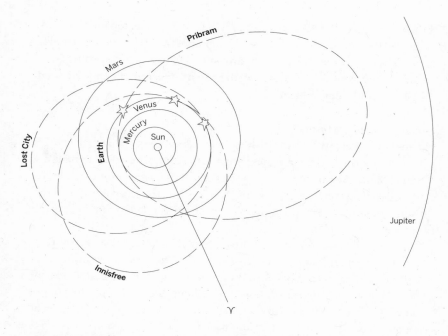

Fig. 10.4. The orbits of the Pribram, Lost City, and Innisfree meteorites are shown as seen from above the inner solar system. The direction to the first point of Aries is also indicated (bottom).[7]

Much of the scientific interest in Innisfree centres around the fact that it was only the third meteorite after Pribram (Czechoslovakia, 1959) and Lost City (USA, 1970) for which two-station photography was available, which permitted its pre-atmospheric orbit to be accurately calculated. The Innisfree meteoroid followed an elliptical orbit (Fig. 10.4) with a period of 2.561 years; an *aphelion* distance, or farthest point from the Sun, of 2.758 *astronomical units* (AU) and a *perihelion* distance, or closest point to the Sun, of 0.986 AU—one AU being equal to the average distance between the Earth and Sun, or about 150 million km. Therefore, the orbit, ranged from very near the orbit of Earth to well beyond the orbit of Mars to the inner fringe of the asteroid belt. The Innisfree object passed through perihelion for the last time just 40 hours before its encounter with the Earth.[8] Exactly three years after the fall of the Innisfree meteorite, a fireball appeared over north-central Saskatchewan, and when Ian Halliday calculated its path from MORP camera data, he found the orbital elements to be essentially identical to those of Innisfree. Halliday postulated that at least one meteorite should have fallen from the second fireball, possibly near the town of Ridgedale, Saskatchewan, and he planned a search for it. He further suggested that, in a manner similar to that proposed for the origin of cometary meteor showers, a collision between two objects had created a stream of debris in an Earth-crossing orbit—including the parent objects of the Innisfree and the putative Ridgedale meteorites.[9]

Mineralogy and Textures

The Innisfree meteorite was initially classified as an L4–5 or olivine-hypersthene chrondrite.[10] The main mass is in the Geological Survey of Canada's National Meteorite Collection in Ottawa. Parts of fragments numbers 1 and 2, plus the whole of number 5 (Fig. 10.5) are in the meteorite collection of the University of Alberta. Small samples have been distributed to a number of laboratories in several countries.

The mineralogy of Innisfree was first reported in detail by D.G.W. Smith (University of Alberta) some three years after the meteorite was recovered.[11] From his examination of a polished thin section (Fig. 10.6), he described it as a "possibly *genomict, brecciated* chondrite." Chondrules and chondrule fragments ranging in size from nearly 2 mm downwards are abundant, making up as much as 50% of the meteorite. Many different varieties of chondrules are present, including excentroradial, various other radiating and feathery types, porphyritic and barred olivine chondrules. The matrix material ranges in size from 0.1 mm downwards. Olivine and pyroxene are the predominant silicate minerals; minor plagioclase, which is well crystallized and not maskelynite, is present in the matrix. Opaque minerals make up less than 10% of the meteorite, with troilite more abundant than metal grains. Smith observed in the thin section of Innisfree two brecciated regions of different petrologic types (5 and 6) separated by a thin vein of opaque material. This is made up of glass that includes extremely small droplets and films of metal and sulphide. Also present in the immediate region of the boundary are occasional grains of troilite. Apart from minor strain effects in these grains, the indications of shock extend for no more than 100 μm from the boundary.

Smith also analyzed the thin section using an electron microprobe fitted with both energy-dispersive and wavelength-dispersive spectrometers. He analyzed 11 olivine grains, finding an average composition of $Fa_{2.71}$; this is at the low end of the range of values associated with the LL group of chondrites. Low-calcium pyroxene grains also showed an average ratio near the low end of the range for LL chondrites; high-calcium pyroxenes showed an average composition more similar to the LL average. Chromite was also found; the two common meteoritic phosphates, merrillite (=whitlockite), and chlorapatite are present;

Fig. 10.5. A painted cast of Innisfree #5.

troilite was the only sulphide identified. The metallic phases of the meteorite were investigated at about 500 analytical points, revealing a wide range of compositions; the overall average is Fe (69.80 %), Ni (29.66 %), Co (0.54 %) by weight. Smith commented that: "the apparent absence of very high Co values in metal grains from Innisfree could reflect the fact that…no attempt was made to measure Co contents at the extreme edges of grains. It is possible, therefore, that high values do exist…."[12] Kallemeyn et al. (see below) subsequently found significantly higher values.

Bulk Chemistry and Trace Elements
G.W. Kallemeyn, A.E. Rubin, D. Wang, and J.T. Wasson (University of California, Los Angeles) measured the concentrations of 26 elements in the Innisfree meteorite and 65 other ordinary chondrites using instrumental neutron activation analysis (INAA) of bulk powdered samples (Table 10.1).[13] Since a major goal of their research was to define the limits of the population of ordinary chondrites, they purposely included meteorites that had previously been reported

Fig. 10.6. A polished thin section of the Innisfree meteorite seen in transmitted light. Note the different textures of the lower (type 5) and upper (type 6) parts, which are separated by a very narrow opaque zone of shocked material. The large white area in the upper left is a hole in the section.

as having some anomalous property; Innisfree was included because it was a recent, photographed fall (the only one in their study). Kallemeyn et al. also prepared polished thin sections of the meteorites, which they examined microscopically in transmitted and reflected light. They made a special effort to identify xenoliths (foreign inclusions), shock veins, chondritic or impact-melt clasts, anomalous metal or troilite abundances, and unusually large chondrites. They made electron microprobe analyses of the fayalite content of olivine crystals and the cobalt content of kamacite in each meteorite. Their olivine and kamacite measurements, substantiated by the INAA elemental abundance data, caused them to reclassifiy Innisfree as a L5 chondrite, notwithstanding the LL5 classification originally assigned by D.G.W. Smith and, subsequently, by Graham et al.[14]

In a series of three papers published over 14 years, Alan E. Rubin (University of California, Los Angeles) reported his investigations of intergroup and intragroup relationships, metallic copper (Cu) content, and the shock and thermal histories of ordinary chondrites. In the first (1990), Rubin reported his high-precision electron microprobe analyses of olivine and kamacite in a large suite of ordinary chondrites including Innisfree, Bruderheim, and Peace River.[15] He hoped, among other things, to more rigourously define the compositional ranges of these minerals for each ordinary chondrite group; identify those ordinary chondrites as fragmental breccias that contain some olivine and/or kamacite grains with aberrant compositions; and, identify anomalous ordinary chondrites whose olivine and/or kamacite compositions lie outside the established ranges and hence may not belong to the three main ordinary chondrite groups. The olivine and kamacite values of Innisfree, when compared with the ranges Rubin established for the three principal groups of ordinary chondrites (H, L, and LL), clearly place it within the L group. In agreement with Kallemeyn et al., he listed Innisfree as a L5. However, although the latter identified Innisfree as containing aberrant kamacite grains, Rubin did not; instead, he listed Bruderheim. He suggested that more than three ordinary chondrite parent bodies were originally formed, but only the H, L, and LL bodies, or their collisional remnants, are now in "efficient meteorite-yielding orbits."[16]

In a later paper, Rubin was interested in metallic Cu because, although it was first discovered in a meteorite in 1919, there were few detailed studies of this phase relative to those of other native metals (indeed, Cu was not among the 26 elements in the Kallemeyn study).[17] He examined polished thin sections of over 100 ordinary chondrites in reflected light with a petrographic microscope. In freshly polished sections, metallic Cu is pale rose in colour; in older sections, the colour is copper-red and often has dark spots due to tarnishing. In Innisfree, Rubin found Cu in the interface between kamacite and troilite grains and also in the troilite-silicate interface. To study the correlation between metallic Cu abundance and shock stage, he first classified the 106 ordinary chondrites using Stöffler's shock stage classification scheme (q.v.); Innisfree was classified as S2, very weakly shocked. He found that although the *amount* of metallic Cu found in ordinary chondrites is independent of shock stage, ordinary chondrites with a relatively large number of occurrences of individual grains (what he termed "occurrence abundance") of metallic Cu "have a tendency to have experienced moderately high degrees of shock."[18]

Table 10.1. Major, minor and trace elements in the Innisfree meteorite.

Method	Na (mg)	Mg (mg)	Al (mg)	K (µg)	Ca (mg)	Sc (µg)	V (µg)	Cr (mg)	Mn (mg)	Fe (mg)	Co (µg)	Ni (mg)	Zn (µg)	Ga (µg)	As (µg)
INAA	7.13	152	12.3	878	13.0	8.50	79.0	3.94	2.61	222	643	12.0	63.0	5.80	1.66
PGNAA															

Note: Concentrations are given in units per gram of meteorite. Method abbreviations are as follows: INAA, instrumental neutron activation analysis; PGNAA, prompt gamma neutron activation analysis.

Rubin (2003) found that chromite ($FeCr_2O_4$), a common oxide in ordinary chondrites, can also be used as a shock indicator.[19] Plagioclase feldspar is compressed and melted relatively easily by the shock of impact and the heat from shock-melted plagioclase causes adjacent chromite grains to melt which, upon cooling and recrystallizing, form "chromite-plagioclase assemblages." These assemblages, ranging in size from 20 to 380 µm, are most common in ordinary chondrites of shock stage S6, occur in nearly every shock stage S3–S5 ordinary chondrite, and also occur in some very weakly shocked S2 ordinary chondrites, including Innisfree.

The abundance of the light element boron in meteorites can shed information on the nature of the pre-solar system nebula and its condensation history. For a number of reasons, it is one of the most difficult elements to study in meteorites. Boron (B) is present in meteorites at levels typically below 1 ppm, near the sensitivity limit of many instruments, so even the slightest terrestrial contamination from water, detergents and soap, or even laboratory glassware, can skew analyses. In 1994, Mingzhe Zhai and Denis M. Shaw (McMaster University) used prompt gamma neutron activation analysis (PGNAA)— together with thoroughgoing contamination-avoidance measures—to analyse the B content of 36 meteorites, including Innisfree, Abee, Bruderheim, and Peace River.[20] They began by requesting, from participating museum curators, interior parts of chosen falls, never touched by water or any other possible sources of B contamination. After preparation, samples were analyzed in the McMaster Nuclear Reactor in Canada and the National Institute of Standards and Technology nuclear reactor in the United States. PGNAA is a good method for B analysis in ordinary rocks, but meteorites present a special problem with their high nickel content: Ni shows a peak in its spectrum at 465 keV, near enough to the B peak at 478 keV to cause background interference. Zhai and Shaw reported the average B concentration, calculated from the data from both reactors, of the Innisfree meteorite as 0.56 ppm. Zhai and Shaw subsequently collaborated with Eizo Nakamura and Toshio Nakano (Okayama University,

Table 10.1. *(continued)*

Method	Se (µg)	Sb (ng)	La (ng)	Sm (ng)	Eu (ng)	Yb (ng)	Lu (ng)	Os (ng)	Ir (ng)	Au (ng)	B (ng)	Reference
INAA	9.7	50	318	202	82	236	34.5	549	513	158		Kallemeyn et al. 1989
PGNAA											560	Zhai and Shaw 1994

Japan) to measure the isotopic ratio $^{11}B/^{10}B$ in nine lunar soil and rock samples, plus most of the original suite of meteorites. Although the $^{11}B/^{10}B$ ratio was essentially the same in all samples, the ratio in Innisfree (4.02322) was the second lowest measured.[21]

Isotopic Chronology

Researchers are always keen to recover a fallen meteorite quickly so that the weak radioactivity produced in the meteorite in space by cosmic rays can be measured. Some of the radioactive isotopes so produced have short half-lives and must be analyzed as soon as possible. For this reason, Halliday wasted no time in taking the meteorite to the Battelle Pacific Northwest Laboratories in Richland, Washington; on February 19, just two weeks after the fall of the Innisfree meteorite, L.A. Rancitelli and J.C. Laul at Batelle began analyzing it using nondestructive gamma-ray techniques.[22] They measured the cosmogenic (cosmic-ray-produced) radionuclides ^7Be, ^{22}Na, ^{26}Al, ^{46}Sc, ^{48}V, ^{51}Cr, ^{54}Mn, ^{57}Co, ^{58}Co, and ^{60}Co as well as the primordial radionuclides of K, U, and Th. The meteorite had an ^{26}Al content of 81 dpm/kg (disintegrations per minute/kilogram)—higher than expected for its chemical composition—suggesting that Innisfree has had an unusually high cosmic ray exposure for the past several million years. In general, though, the abundances of 13 rare earths; Ba, Sr, and U; and other trace elements as measured by radiochemical neutron activation analysis are the same as the average chondritic abundance, confirming that Innisfree is an ordinary chondrite.

Two years after the Battelle study of radionuclides in the meteorite, A.K. Lavrukhina and V.D. Gorin (Academy of Sciences of the USSR, Moscow) reported similar results.[23] G. Heusser, W. Hampel, T. Kirsten, and O.A. Schaeffer (Max-Planck-Institut für Kernphysik, Heidelberg, Germany) measured several cosmogenic radionuclides by gamma and beta spectroscopy. Pre-atmospheric sizes of the meteorites and the shielding depths of the investigated samples were estimated by a comparison of the measured ^{22}Na, ^{54}Mn, and ^{60}Co with calculations by other researchers. They estimated a diameter of 10–40 cm for the pre-atmospheric Innisfree, with a shielding depth of less than 10 cm. Heusser and his colleagues also analyzed the stable rare gas isotopes in the meteorite by mass spectrometry. It will be recalled that Rancitelli and Laul had suggested, based upon Innisfree's unexpectedly high ^{26}Al content, that it had received an unusually high cosmic ray exposure. Heusser et al. came to the same conclusion after they measured a lower than expected ^{37}Ar/^{39}Ar ratio. They estimated the cosmic-ray exposure age of the Innisfree meteorite, from the production rates (corrected for shielding effects) of ^{3}He, ^{21}Ne and ^{38}Ar, as 28 ± 3 Ma. Their potassium-argon (K-Ar) dating of Innisfree yielded an age of 4.1 ± 0.3 Ga. Finally they reported that Innisfree contains solar type trapped gases, "which is very rare among L-chondrites."[24]

The late O. Müller and his colleagues W. Hampel, T. Kirsten and G.F. Herzog, also at the Max-Planck-Institut für Kernphysik, followed a similar methodology and estimated that cosmic ray intensity had varied by less than 20% in the last three or four half-lives of ^{53}Mn (36–48 Ma), with a possible recent (≤ 2.5 Ma) increase.[25]

Contemporaneously with the work of G. Heusser's group, a team led by J.N. Goswami (Physical Research Laboratory, Ahmedabad, India) carried out rare-gas and particle track studies on samples of Innisfree (a total of five samples from fragments nos. 2, 3, 4, and 6) provided by H.R. Stacy (Geological Survey of Canada) and R.E. Folinsbee.[26] Using a conventional Reynold's type glass mass spectrometer Goswami and his co-workers measured the ^{3}He, ^{21}Ne, and ^{38}Ar abundances in three of the samples; based on the observed (^{3}He/^{21}Ne) and (^{22}Ne/^{21}Ne) ratios, they estimated a cosmic-ray exposure age of ~26 Ma for the Innisfree meteorite. This agrees well with the value of 28 ± 3 Ma obtained by Heusser et al. They estimated the pre-atmospheric mass of the Innisfree meteorite as ~400 kg, with an uncertainity of a factor of ~2. Goswami and his colleagues carried out track analyses in olivine grains, which is the dominant mineral species in Innisfree, from all five samples and in pyroxene grains from two samples. From the track data they estimated a shielding depth of ~20 cm and a pre-atmospheric radius of 28–32 cm.

Innisfree played a role in the development of the thermoluminescence method for dating meteorite finds. C.L. Melcher (Washington University in St. Louis) and D.W. Sears (University of Birmingham) measured the induced ther-moluminescence (TL) in the Innisfree meteorite and nine other chondrites.[27] Powdered samples were prepared from interior portions of the meteorite specimens to avoid heat-altered material from near the fusion crusts. This was important because the heat of atmospheric passage completely erases the natural TL near the surface of a meteorite. The natural TL would then be rebuilt at an unknown rate by terrestrial radiation, greatly complicating age determination. Instead, Melcher and Sears erased the natural TL from the samples in a controlled manner by heating them in an oxygen-free argon atmosphere. They then irradiated the samples and measured the remaining TL after 20 and 68 hours. Melcher and Sears were able to show that the rate of decay is slower than exponential and also that it varies with terrestrial age. Young, recently fallen meteorites decay somewhat faster than older meteorites. Melcher and Sears concluded that their findings result "in a simplification in the use of TL to date meteorite finds by eliminating the need for determining the decay characteristics of each meteorite."[28]

Summary

The Innisfree meteorite is an ordinary chondrite and, in most respects, unremarkable. Much of the scientific interest in Innisfree centres around the fact that it was only the third meteorite to have its pre-atmospheric orbit accurately calculated. Also, it was used in an important study of the use of thermoluminescence to simplify the dating of meteorites.

11

The Peace River Meteorite

History

A large bolide can be a truly spectacular sight as Emil Grigoleit, a farmer living near Hines Creek in northwestern Alberta, attested. Mr. Grigoleit's description of the 31 March 1963 Peace River fireball is one of the best ever given:

> In regard to the explosion heard on Sunday morning,...we heard it at approximately 4:40 a.m. It was preceded by a light, which we saw through our window of the size of a flashlight, which grew rapidly to fill the room with an orange glow. Then came a flash so brilliant that you could not see anything and it lit up everything bright as day. Then came the explosion after the light faded out, which shook the house with such force that the windows and dishes rattled. The whole affair did not last more than ten seconds at the most from the time the light was first seen until the explosion subsided.[1]

Mr. Grigoleit's account continues: "the glow of the path of the object seemed to have met the earth at a point south-south-east of us. We did not see the object strike the earth as we are in a valley and trees cut off any view we would have had of it." The flash was visible for over 160 km, and the detonations, likened to sonic booms, were heard over a 10,000 km² area. It was now that arrangements that were made as a result of the Bruderheim meteorite shower three years earlier came into play. A reporting system set up by the newly created Associate Committee on Meteorites (ACOM) of the National Research Council of Canada resulted in the gathering of a substantial amount of sighting data by the Royal Canadian Air Force (RCAF) and the Royal Canadian Mounted Police (RCMP). These reports were collated by a RCAF officer and telephoned or wired to Professor Folinsbee at the University of Alberta.[2]

It was apparent that a detonating bolide had travelled on a generally west to east path across northern Alberta and that chances of meteorite recovery were good. Dr. L.A. Bayrock and Len Hills, a graduate student in geology and former resident of the Peace River district, drove 500 km to Peace River to interview eyewitnesses at the point of their observations. They had observers, at the exact spot of sighting, point a stick in the directions sighted and the angles were measured with a Brunton compass. This was necessary if a possible fall area was to be accurately pinpointed in that vast and sparsely populated country. The following reports, presented in chronological order of data gathering, are some of the most important eyewitness accounts used to delineate the fall area (Fig. 11.1).

Constable E.W. Lowe of the RCMP was travelling north on Highway 2, two miles (3.2 km) north of Nampa, when the interior of the car and the ground outside were lit up like from sheet lightning and radio station CJCA reception was interrupted by cutout. Fifteen seconds later he observed directly west a dark object, "which disappeared after forming a massive flame...."

Mr. Halonen, travelling north on Highway 2 on high ground 11 km south of Valleyview, had a good, although 140 km distant, view of the event. He saw the bolide break up into two pieces; the smaller appeared to burn up, and the larger disappeared over the horizon.

Mr. Lloyd Leonard in the town of Peace River saw an orange fireball the size of several full moons descending at a very steep angle from south to north. This exploded to the southwest at an elevation of 19° and disappeared leaving behind a large number of fragments, which fell to the ground. About one minute afterwards, he heard a low rumbling noise which appeared to come from all directions.

Mr. Fred St. Germain of Shaftesbury Settlement stepped out of his house and heard a low hissing sound. Looking at the sky, St. Germain saw a bolide, the size of the full moon, travelling from south to north; it disappeared behind the river bank.

Mrs. Ragelson, in Peace River, was walking eastwards, saw the sky brighten, turned and saw a reddish bolide about one-third the size of a full moon descending from south to north at an angle of 40° to the horizontal; it disappeared behind the river bank.

Mr. R. Proctor, in Bluesky, saw a flash in the sky and then a yellow to reddish fireball about one-tenth the size of the full moon, and at the same time heard a low rumbling sound. He first saw the fireball at 15° elevation in a direction N37°E. It was travelling west to east and disappeared at 9° elevation at N57°E.

Mr. Peter Karpiak, Warrensville, was opening a barn door when he saw the whole sky brighten up. Turning around, he saw a fireball of very deep red colour about "1 foot" (0.3 m) in diameter. The bolide was first sighted at S38°W and it disappeared at S33°W.

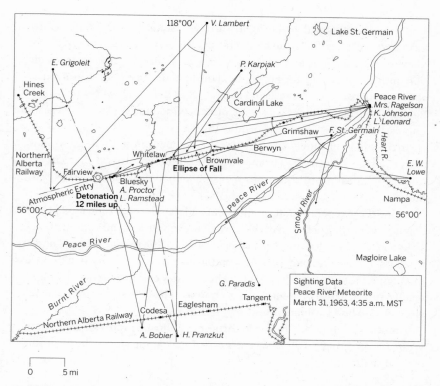

Fig. 11.1. Bayrock's revised plot of all known direct observations of the Peace River bolide, placing the fall ellipse farther southwest.

Mr. V. Lambert of Smith Falls was driving a little west of due south on a winding logging road when he saw a bolide travel across the path of his car from west to east and noted the explosion. The bolide, small at first, became larger and larger, and when it exploded it was at least the size of the full moon. It broke into many pieces of varying size. Mr. Lambert thought the explosion occurred 24–32 km away. His sense of direction and guess at distance proved to be very precise, and his account of the explosion best fits the actual fall pattern.

Mr. Ken Johnson, Peace River, first saw a bright flash, and on looking westward saw a bolide the size of the full moon descending at a very steep angle (45°) from south to north. It exploded and broke into two large fragments and many small ones. The bigger of the main fragments fell northwards at an angle of 56° to the horizontal, and the other plunged to the south at an angle of 75°.

Alfred Bobier, the 18-year-old son of Mr. C.E. Bobier of Belloy, was out near the barn looking for newborn calves and lambs at about 4:35 a.m. when the whole sky was lit up with a blue light. Turning first north, then south, he saw an orange ball of fire followed by several small ones which were not so bright. These objects first appeared in the northwest, travelled in a northeasterly direction until a little east of due north where the light disappeared. He walked about 100 m back to the house and had just closed the door when the two Bobiers, father and son, heard a loud, prolonged sound, which was all the more impressive because "our walls are quite thick."

In the village of Tangent, Mr. Gustav Paradis saw, through a north window, a large red fireball travelling from west to east. It broke into two, one part continuing in the same direction, the other smaller part descending vertically to the ground.[3]

Using these reports, Dr. Bayrock calculated that the explosion took place about 5 km north-northwest of the town of Brownvale at an altitude of about 13 km. From this calculation, possible fall areas of the two large pieces and the small fragments were plotted. A Cessna 180 aircraft was then used in the search for fragments; however, because of the heavy snow cover and fresh drifting, results were negative. A preliminary ground search of the snow-covered fields also failed to find any meteorites.

Ground searching was begun in earnest by Folinsbee and a student, John Westgate, on April 23. Folinsbee and Bayrock gave some idea of the magnitude of the task that lay ahead:

Searching in stubble fields for the Bruderheim meteorite had shown that even large pieces could be missed unless fields were systematically swept, with not more than a 25-foot [8 m] interval between searchers. To cover one square mile effectively requires 210 man-miles [336 person-kilometres] of traversing. For Peace River the likely area covered at least 70 square miles [180 km^2], and the problem of 15,000 miles [24,000 km] of traversing was a formidable one.[4]

During the traverse of the first field searched in the Brownvale area, John Westgate discovered Peace River #1, an 8 kg stone that had fragmented on impact (Fig. 11.2). One fragment of this first meteorite had apparently been found by a coyote. Perhaps disappointed that the black object was not edible, the animal defecated on it and departed. Folinsbee and Bayrock wryly speculated on the consequences of having this fragment's chemical composition analyzed.

Permission to continue the ground search was now obtained from various landowners in the district. They were shown the meteorite and encouraged by

Fig. 11.2. John Westgate holding a fragment of Peace River #1.

an offer of five dollars a pound to do some searching themselves. Searching by mixed parties of University of Alberta and Alberta Research Council person-nel and area residents continued during May without success, but important contacts were made with district farmers. Further demonstrations of meteor-ites were made to farmers who were beginning their own systematic sweeps of fallow and stubble fields in the course of spring seeding operations. Identifying meteorites was complicated by the widespread presence of black chert pebbles with an uncanny resemblance to the real thing. Such imposters have been termed "meteorwrongs" by meteoriticists.

Early in June, two additional finds, Peace River #2, a fine, almost complete museum-quality specimen weighing 9.6 kg, and Peace River #3, a small partial individual weighing 355 g, were discovered by district farmers R. Gardiner and O. Vasrud.[5]

Cliff Davies then recovered the two largest individuals, Peace River #4 (16.5 kg) and #5 (11.3 kg), which had been underwater, thus eluding earlier searches. During seeding operations, they had been picked up from the slough margin as glacial erratics and tossed over the fence. There they lay, until the continuing meteorite publicity prompted a further check. The two fragments are partial individuals, with surfaces oxidized by immersion in water and subsequent drying in air, and their exact original positions in the field cannot be precisely ascertained. R.S. Johnson and his son discovered two final, small pieces, the 1.2 kg Peace River #6 and the 2.3 kg Peace River #7.

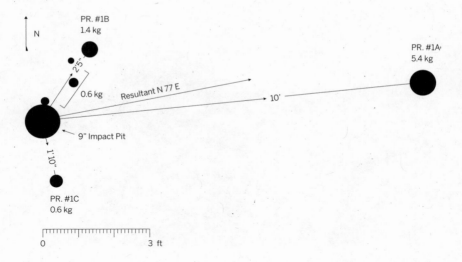

Fig. 11.3. Peace River #1 retained a small component of its cosmic trajectory at the time of impact, as indicated by the distribution of fragments.

Science

Fall Phenomena and Astronomical Interpretation

Careful mapping of the fragments of Peace River #1 relative to the point of impact showed that this individual had been falling under the influence of gravity, while retaining a small component of its original cosmic trajectory with a resultant orientation of N77°E (Fig. 11.3). This is very close to the revised trajectory (N73.5°E) plotted from sighting data and subsequently confirmed by the long axis of the ellipse of fall (Fig. 11.4).[6]

Peace River #1 was recovered 25 days after its fall; a substantial part of the interesting argon (Ar) radioisotope [37]Ar (half-life = 35 days) had already decayed, and it had started to rust (Fig. 11.5). When Mr. E.A. Mahood, the landowner, generously donated the meteorite to the University of Alberta for scientific purposes, distribution of samples to other investigators began immediately, and a note announcing its recovery was sent to the *Meteoritical Bulletin*.[7] So keen was Folinsbee to promote the scientific study of Alberta meteorites that he concluded an early paper with what is tantamount to a sales pitch: "Small amounts of the Peace River meteorite are available for investigators interested in any phase of meteorite analysis, and Bruderheim is still available in quantity."[8]

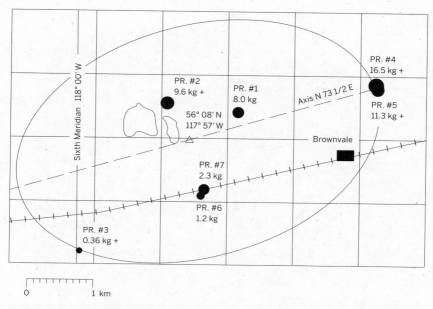

Fig. 11.4. The Peace River meteorite ellipse of fall. Note that the heavier fragments are congregated at the eastern apex of the ellipse and compare the orientation of the long axis with Fig. 11.3.

With the information obtained from these seven meteorites, weighing a total of 49.3 kg, Folinsbee and Bayrock reconstructed the ellipse of fall (Fig. 11.4).[9]

Mineralogy and Textures

The widespread distribution of Peace River specimens, after recovery of the first fragment in 1963, resulted in the publication of several papers the following year. H. Baadsgaard and R.E. Folinsbee (University of Alberta) provided a list of the researchers and institutions that received samples and the nature of their investigations in their preliminary chemical and mineralogical description of the Peace River meteorite.[10] Baadsgaard and Folinsbee found that the bulk chemical composition of Peace River is very similar to both Bruderheim and Kyushu, a Japanese meteorite; all are olivine-hypersthene chondrites. Peace River consists mostly of silicon dioxide (39.81 %), magnesium oxide (24.78 %), iron oxide (12.91%), and metallic iron (8.18%); the metallic nickel (Ni) content is 1.32%. The density of Peace River is 3.60 ± 0.01, a value close to the average density of all chondrites, 3.58. Likewise, the atomic abundances of magnesium (Mg), aluminum (Al), silicon (Si), calcium (Ca),

Fig. 11.5. Peace River meteorite sample #11547/PR1 (University of Alberta) showing oxidation (rusting), the result of over three weeks' exposure to water and air.

iron (Fe), Ni, sodium (Na), and potassium (K) are very close to the averages for L chondrites.[11]

The Peace River chondrite consists largely of olivine, orthopyroxene, plagioclase, kamacite, taenite, and troilite. The plagioclase has been extensively shock-metamorphosed to its glass-like equivalent, maskelynite. The meteorite is distinctly recrystallized and chondrules are poorly preserved, particularly on their margins. In addition to these indications of shock, Peace River contains thin, black, sulphide-rich shock veins similar to those found in other very strongly shocked meteorites (Fig. 11.6). Typically, within rounded fragments found in these veins, the constituent grains of olivine and orthopyroxene have been transformed into their high-density *polymorphs*: *ringwoodite* and *majorite*, respectively.[12] One larger shock vein, several millimetres thick (Fig. 11.7), studied by G.D. Price, and A. Putnis (Cambridge University), and

Fig. 11.6. The recrystallized structure of the Peace River meteorite, with shock veins, seen in transmitted cross-polarized light. (30× magnification). Field of view 4.9 mm. For colour image see p. 289.

0   1   cm

Fig. 11.7. Peace River meteorite sample #11550/PR4 (University of Alberta) showing the thick shock melt vein in which wadsleyite was discovered.

The Peace River Meteorite   123

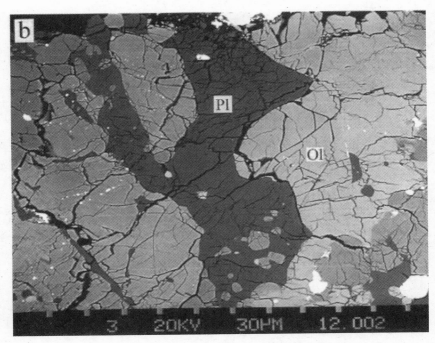

Fig. 11.8. Plagioclase (Pl) in the Peace River meteorite, showing a complex network of cleavages, cracks, and fractures; Ol is olivine. Spacing of scale bars at bottom is 30 μm.

D.G.W. Smith (University of Alberta) differs from those described in other meteorites, in that it contains significant quantities of a new mineral, the naturally occurring β-phase polymorph of olivine, discovered by Price and his co-workers.[13]

In a subsequent paper, Price, Putnis, Smith, and S.O. Agrell (Cambridge University) named the mineral *wadsleyite*, in honour of the late Arthur David Wadsley, a renowned Australian crystallographer. The type specimen, #11550/ PR4, is preserved in the meteorite collection of the Department of Earth and Atmospheric Sciences, University of Alberta. Price and his co-workers examined wadsleyite crystals by transmission electron microscopy and electron microprobe. They also obtained X-ray diffraction data from powdered wadsleyite crystals hand-picked from the vein in the Peace River meteorite.[14] Price later provided information on the nature and significance of stacking faults in the crystal structure of wadsleyite.[15]

Ming Chen (Chinese Academy of Sciences) and Ahmed El Goresy (Max-Planck-Institut für Chemie, Mainz, Germany) presented evidence that

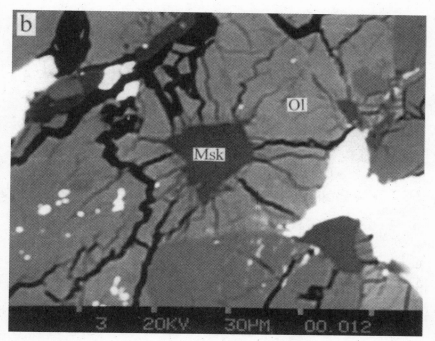

Fig. 11.9. Maskelynite grains (Msk) in the Peace River meteorite are smooth and unbroken in comparison to abundant fractures and cracks in neighbouring olivine (Ol). Spacing of scale bars at bottom is 30 μm.

maskelynite in meteorites is not diaplectic glass, formed without melting by shock-induced solid-state transformation of plagioclase but, rather, is formed by shock-induced melting and quenching of the dense melt at high pressure.[16] A common feature of *diaplectic* feldspar in shocked meteorites like Peace River is the abundant occurrence of cleavages, cracks, and shock-induced fractures (Fig. 11.8).

In comparison, Chen and El Goresy showed that maskelynite is smooth and lacks cleavages, cracks, and shock-induced fractures. Instead, maskelynite grains are surrounded by radiating and open expansion cracks, emerging from their surfaces and shattering the surrounding minerals (Fig. 11.9). These radiating expansion cracks are not filled with maskelynite; this indicates that the plagioclase was melted and quenched to a dense glass at high pressure and that, when the pressure was released, the glass expanded in volume and cracked the surrounding silicates.

In Peace River, maskelynite is abundant near the shock veins. Abundance of unfractured maskelynite decreases with increasing distance from the

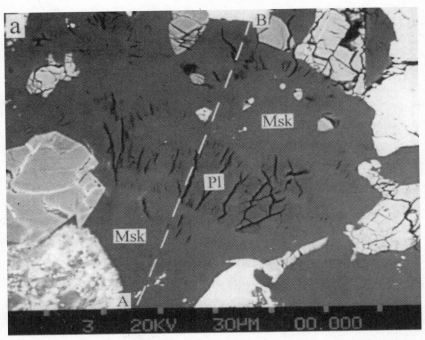

Fig. 11.10. A composite grain in the Peace River meteorite consisting of smooth areas of maskelynite (Msk) and relicts of plagioclase fragments (Pl) characterized by fractures and cracks. The line A–B is the compositional profile analyzed by electron microprobe. Spacing of scale bars at bottom is 30 μm. Compare with Fig. 11.9.

shock veins. A composite maskelynite grain located about 200 μm from a vein was found to contain residues of rounded crystalline plagioclase fragments with fractures and cracks terminating at contact with the unfractured regions (Fig. 11.10). This indicates that the plagioclase was incompletely melted, and the unmelted fragments were "frozen" in place. With increasing distance from the vein, maskelynite is gradually replaced by diaplectic feldspar. In Peace River, maskelynite occurs in 100–500 μm wide areas on both sides of shock veins.

Electron microprobe analysis across the plagioclase-maskelynite interface revealed that the latter is substantially enriched (~300%) in K relative to the diaplectic plagioclase.[17]

W.R. Van Schmus (University of Kansas) and P.H. Ribbe (Virginia Polytechnic Institute) used electron microprobe analysis and X-ray diffraction techniques to determine the chemical compositions and structural states of feldspar, a mineral consisting of silicates of aluminum with varying amounts of Na, K, and Ca. in some 30 chondrites. They found that the feldspar in Peace River consisted of 83.8% albite (Na-rich feldspar); 10.6% anorthite (Ca-rich); and 5.6% orthoclase (K-rich).[18]

A study of *melt* (small, closed bodies of brown glass) in L chondrites led R.T. Dodd (State University of New York, Stony Brook) and E. Jarosewich (U.S. National Museum, Washington, DC) to propose a refined shock classification scheme for equilibrated chondrites. They established that melt pockets result from shock-melting *in situ*, unlike shock melt veins, whose filling needs not be of local origin; that their appearance coincides with and may be related to loss of ^{40}Ar; and that the presence of melt pockets is a useful new criterion for shock intensity. Dodd and Jarosewich distinguished six facies of shock (a, b, c, d, e, and f) ranging from "olivine fractured, plagioclase wholly or chiefly undeformed" to "olivine recrystallized, plagioclase maskelynite;" the Peace River chondrite is assessed as facies d. They propose a concise shorthand expression (e.g., L6d) for the primary (chemical *group*), secondary (petrologic *type*), and tertiary (shock *facies*) properties of any equilibrated chondrite.[19]

Bulk Chemistry
R.A. Schmitt, G.G. Goles, R.H. Smith, and T.W. Osborn (Gulf General Atomic and the University of California, San Diego) limited their study of elemental abundances in 103 chondrites, including Peace River, to those elements that can be studied by instrumental neutron activation analysis (INAA). They presented data on the abundance of eight elements: Na, Al, scandium (Sc), chromium (Cr), manganese (Mn), Fe, cobalt (Co), and copper (Cu) (Tables 11.1 and 11.2a).[20]

James R. Vogt and William D. Ehmann (University of Kentucky), another research team to receive a sample of Peace River #1 from R.E. Folinsbee, determined the silicon abundance in more than 100 chondrites and achondrites.[21] They used fast neutron activation analysis (NAA), requiring less than 10 minutes per analysis with a precision, based on four separate analyses of each meteorite, better than ± 3% and an accuracy probably better than ± 5%.[22]

Minor and Trace Elements
H.W. Taylor (University of Alberta) received a 1.65 kg sample of Peace River and a 1.73 kg piece of Bruderheim for a study of their radioactivity.[23] The radioactivity created in a meteorite by cosmic ray bombardment can be analyzed by measuring the gamma rays emitted by the meteorite as the radioactive atoms decay with a scintillation counter. This method has the advantage that the meteorite is not damaged or altered in any way. However,

Table 11.1. Major and minor elements in the Peace River meteorite.

Method	Si (mg)	Al (mg)	Fe (mg)	Ca (mg)	Na (mg)	Mg (mg)	Mn (mg)	Ni (mg)	Cr (mg)	K (µg)	Reference
NAA	174										Vogt and Ehmann 1965
Wet and dry chemical analyses			81.8a					13.2			Baadsgaard et al. 1964
INAA		11.2	175		6.3		2.5		4.9		Schmitt et al. 1972
NAA								12.3			Müller et al. 1971
NAA				8.2	7.8	177				870	Shima and Honda 1967b

Note: Concentrations are given in units per gram of meteorite. Method abbreviations are as follows: NAA, neutron activation analysis; INAA, instrumental neutron activation analysis.

aFree (metallic) iron.

meteorites are only weakly radioactive so a large (>1 kg) sample is needed to create a detectable level of radiation. In addition, Taylor placed each meteorite and the scintillation counter inside a light-proof and lead-shielded box; even so, for both the Bruderheim and Peace River samples, the counting rate was only about 20% above the background counting rate ("noise"). Taylor measured Bruderheim first to confirm previous results by other investigators, then he tested Peace River. His measurements indicated the presence of the radioisotopes [22]Na, [54]Mn, [40]K, and [26]Al in the Peace River meteorite. He found no evidence of other radioisotopes, such as [46]Sc, [56]Co, [58]Co, and ^{60}Co that have been observed in other meteorites, possibly due to the uncertainty caused by subtracting such a large background.

Julian P. Shedlovsky, Philip J. Cressy, and Truman P. Kohman (Carnegie Institute of Technology) measured the abundances and ratios of a more extensive range of cosmogenic radionuclides, with half-lives from about two weeks to 269 years, in the Peace River and Harleton, Texas chondrites. The elements of interest were separated by various chemical extraction techniques and the radionuclides measured as follows:

[26]Al was determined by coincidence counting of the positron annihilation radiation. Gamma spectrometry was used for measuring [54]Mn and [56,57,58]Co. A gas-filled internal-sample X-ray proportional counter was used to measure [49]V, ^{51}Cr, [53,54]Mn, [56,57,58]Co, and [59]Ni. Low-level beat Geiger counters were used to measure [10]Be, [36]Cl, [44]Sc milked from [44]Ti, [46]Sc, [48]V, and [60]Co. [10]Be, [36]Cl, and ^{60}Co were chemically recycled to constant specific activity, and the decays of [44]Sc, [46]Sc, [48]V, and ^{60}Co were followed. [[22]Na and ^{14}C were also measured].[24]

Shedlovsky and his colleagues reported that their Peace River [46]Sc results seemed "extraordinarily high." They noted that there was nothing in Peace River's chemical composition that would account for this enrichment and suggested that their sample had been "fortuitously situated in a pre-atmospheric body of optimum size" for exposure to a high flux of cosmic radiation, a suggestion, they said, that was supported by the high [10]Be activity in Peace River. The picture they had "is of a pre-atmospheric body not much larger than the 40 kg recovered mass, with our specimen within perhaps 5–10 cm of one surface."[25]

The Shedlovsky et al. Peace River results were included by Philip J. Cressy in a table of similar analyses of several other meteorites (and one artifical satellite, Sputnik 4!). Peace River appears unremarkable, except in sharing a higher than expected [46]Sc/[54]Mn ratio with two other meteorites.[26]

The chemical determination of the six closely associated platinum metals in iron and stony meteorites has always presented difficulties because of the complexities involved in separating these metals from the other elements and from each other. Notwithstanding these problems, J.G. Sen Gupta (Geological Survey of Canada, Ottawa) determined the ruthenium (Ru), rhodium (Rh), palladium (Pd), osmium (Os), iridium (Ir), and platinum (Pt) content of the Peace River and some 28 other meteorites (see Table 11.2b). After decomposition of the meteorite samples by perchloric acid, followed by ion-exchange separation, the elements were measured spectrophotometrically. Of these six elements, only Ru was not detected in Peace River.[27]

Otto Müller, Philip A. Baedecker, and John T. Wasson (University of California, Los Angeles) reported mean bulk concentrations of Ir and Ni in Peace River, measured by NAA.[28] Their Ir value was less than Sen Gupta's and their Ni value was lower than that of Baadsgaard and Folinsbee (1.32 %), no doubt reflecting the problems of chondrite analysis mentioned by Schmitt et al. (see above).

L. Greenland and Gordon G. Goles (University of California, San Diego) determined the concentration in Peace River of Cu using a neutron activation technique employing radiochemical separations; their result was higher than a previous INAA result by other workers (Table 11.2a). Greenland and Goles concluded that "the divergent values are due to sample inhomogeneity rather than to faulty analytical techniques," and that "in view of the large sampling error the conclusions to be drawn from these analyses must be somewhat tentative." They also reported the Zn concentration in Peace River.[29]

The distribution of alkali, alkaline earth, and rare earth elements in three Alberta meteorites—Abee, Bruderheim, and Peace River—was studied by

Table 11.2a. Trace elements in the Peace River meteorite.

Method	Co (µg)	Rb (µg)	Sr (µg)	Sc (µg)	Cu (µg)	Zn (µg)	Ga (µg)	Ge (µg)
INAA	610			9.0	91			
INAA					120	56		
NAA		3.1	10.8					
NAA							5.1	9.9
Spectrophotometry							11.5[a]	126.5[a]
							5.2[b]	0.22[b]
NAA								
Wet and dry chemical analyses	500							
NAA	600							
Nuclear resonance								

Note: Concentrations are given in units per gram of meteorite. Method abbreviations are as follows: NAA, neutron activation analysis; INAA, instrumental neutron activation analysis.

[a] In the metal phase only.

[b] In the silicate phase only.

Masako Shima and Masatake Honda (University of Tokyo). They used a step-wise fractional dissolution method to decompose the various main component minerals successively, followed by NAA. Shima and Honda reported their Peace River results as whole meteorite concentrations of Na, K, rubidium (Rb), Mg, Ca, and strontium (Sr) in parts per million.[30]

Shima and Honda reported rare earth element abundances for Abee and Bruderheim but not for Peace River. Fortunately, Akimasa Masuda and Noboru Nakamura (University of Tokyo) and Tsuyoshi Tanaka (Geological Survey of Japan) determined these for ten chondrites, including Peace River, by mass spectrometric isotope dilution of dissolved meteorite samples. They reported the abundance of lanthanum (La), cerium (Ce), neodymium (Nd), samarium (Sm), europium (Eu), gadolinium (Gd), dysprosium (Dy), erbium (Er), ytterbium (Yb), and lutetium (Lu) in Peace River (Table 12.2b).[31]

The distribution of tungsten (W), molybdenum (Mo), and Co between the metal, silicate, and troilite (sulphide) phases of Peace River and some other chondrites was investigated by Keiko Imamura and Masatake Honda (University of Tokyo) using NAA. They found that, in Peace River, W was heavily concentrated in the metal phase; Mo was almost equally distributed in the metal and troilite phases, with little in the silicate phase; 98% of the Co was found in the metal phase.[32]

The elements gallium (Ga), germanium (Ge), and Ir have been extensively studied in iron meteorites but less so in chondrites. S.N. Tandon and John T.

Table 11.2a. *(continued)*

Method	N (µg)	F (µg)	Mo (µg)	W (ng)	Reference
INAA					Schmitt et al. 1972
INAA					Greenland and Coles 1965
NAA					Shima and Honda 1967a
NAA					Tandon and Wasson 1968
Spectrophotometry					Chou and Cohen 1973
NAA	21				Kothari and Goel 1974
Wet and dry chemical analyses					Baadsgaard et al. 1964
NAA			1.31	160	Imamura and Honda 1976
Nuclear resonance		32			Allen and Clark 1977

Wasson measured these elements plus a fourth, indium (In), in a suite of 26 L chondrites, including Peace River. Tandon and Wasson made replicate neutron activation determinations and found that the range of Ga, Ge, and Ir concentrations in all the meteorites is small, varying by a factor of 1.5–2.0. The most striking aspect of their data is the range in levels of In from only 0.048 ppm in Peace River to 55 ppm in Hallingeberg, a factor of more than 1000. Indium concentrations are strongly correlated with petrologic type: the severely shocked L6 chondrites, such as Peace River, have very low levels of In; the weakly shocked L3 chondrites, such as Hallingeberg, have the highest levels. This suggested to Tandon and Wasson that In is a highly mobile element and probably a highly volatile one. It also suggested that In should be compared with other variables that may reflect the thermal history of a meteorite. The concentrations of primordial ^{36}Ar and ^{132}Xe are also strongly correlated with petrologic grade.[33]

The elemental abundances reported by Tandon and Wasson are bulk compositions; meteorite samples were powdered before testing, with no attempt to separate phases. However, Chen-Lin Chou and A.J. Cohen (University of Pittsburgh) determined Ga and Ge levels in carefully separated metal and silicate phases for Peace River and numerous other chondrites.[34] They reported mean values for Ga in the metal and silicate phases of Peace River of 11.5 ppm and 5.2 ppm, respectively; corresponding values for Ge were 126.5 ppm and 0.22 ppm. Thus, Ge is more concentrated in the metal phase than Ga; given that Peace River's metal phase is only 7.5% of the whole[35], the

Table 11.2b. Trace elements in the Peace River meteorite.

Method	B (ng)	La (ng)	Sm (ng)	Eu (ng)	Yb (ng)	Lu (ng)	Nd (ng)	Gd (ng)	Ce (ng)	Er (ng)	Dy (ng)	Pd (ng)
Spectrophotometry												420
NAA												
Mass spectrometry		412	231.4	86.2	245	38.2	744	309	1052	253	375	
NAA												
PGNAA	1170											
Delayed neutron activation												

Note: Concentrations are given in units per gram of meteorite. Method abbreviations are as follows: NAA, neutron activation analysis; PGNAA, prompt gamma neutron activation analysis.

proportions of Ga and Ge in Tandon and Wasson's bulk composition are accounted for.

Masako Shima and Masatake Honda (University of Tokyo) determined the Ru and Sr levels in the separated mineral components of three Alberta meteorites: Abee, Bruderheim, and Peace River.[36] They removed the magnetic fraction from finely crushed samples with a hand magnet and then treated the non-magnetic fractions with several successive reagents. Rubidium and strontium concentrations were determined by isotope dilution analyses, some of which were confirmed by neutron activation methods. The isotopic composition and ratio of Rb and Sr ($^{87}Sr/^{86}Sr$ and $^{87}Rb/^{86}Sr$) were measured by mass spectrometry. Shima and Honda found that their data for separated minerals were not very useful for a detailed calculation of the Rb-Sr age of Peace River, but their whole meteorite (bulk) data was consistent with an age of 4.5 Ga.[37]

Aware of the difficulties presented by the isotope dilution analysis method described above, G.L. Cumming (University of Alberta) chose to use the delayed neutron method to measure uranium (U) and thorium (Th) in Peace River.[38] This method depends on the fact that some fissionable elements such as U and Th yield some fission products, which decay by emitting delayed neutrons following β-decay. These so-called delayed neutron emitters have half-lives in the range of 1–60 s. Because thermal (slow) neutrons will induce fission in ^{235}U, whereas fast neutrons are required for ^{238}U and ^{232}Th fission, it is necessary to irradiate the sample twice, once by mixed neutrons and again with a cadmium shield around the sample (to absorb the thermal neutrons) to obtain enough information to determine both U and Th.[39]

The noble, or inert, gases (helium, He; neon, Ne; argon, Ar; krypton, Kr; xenon, Xe; and radon, Rn) have been extensively studied in meteorites,

Table 11.2b. (continued)

Method	Os (ng)	Ir (ng)	Pt (ng)	Rh (ng)	Th (ng)	U (ng)	In (ng)	Reference
Spectrophotometry	540	620	530	300				Sen Gupta 1968
NAA		450						Müller et al. 1971
Mass spectrometry								Masuda et al. 1973
NAA		520					48	Tandon and Wasson 1968
PGNAA								Zhai and Shaw 1994
Delayed neutron activation					96	11.8		Cumming 1974

primarily for determining cosmic-ray exposure and gas-retention ages.[40] The more chemically reactive gases such as nitrogen (N) and fluorine (F) have received less attention. One of the difficulties faced in measuring N in meteorites is eliminating sample contamination by the terrestrial atmosphere (78% N). B.K. Kothari and P.S. Goel (Indian Institute of Technology, Kanpur) determined the total N content in several types of meteorites including chondrites by NAA using a reaction in which N atoms absorb neutrons, emit protons, and transmute into carbon atoms.[41] They found the bulk N content of Peace River to be 21 ppm (mean of three samples).[42] Ralph O. Allen and Patrick J. Clark (University of Virginia, Charlottesville) determined the F content of 21 meteorites by proton beam irradiation using a reaction in which F atoms absorb protons, emit alpha particles (helium nuclei), and gamma rays, and transmute into oxygen atoms. They were the first to measure the bulk F content of Peace River, finding a mean value of 32 ppm.[43]

Mingzhe Zhai and Denis M. Shaw (McMaster University) used prompt gamma neutron activation analysis (PGNAA) to determine the average boron (B) concentration of the Peace River meteorite as 1.17 ppm, the highest level among the several L6 chondrites tested.[44] Zhai and Shaw, in collaboration with Eizo Nakamura and Toshio Nakano (Okayama University), later measured the isotopic ratio $^{11}B/^{10}B$ in Peace River as 4.06332, among the highest measured.[45]

Meteoriticists are interested not just in the chemical elements and inorganic compounds, but also the *organic* molecules, to be found in meteorites.[46] In the 1960s, scientists were especially interested in the organic compounds found in meteorites because of the implications for the origin of life on Earth and the existence (or not) of extraterrestrial life. One of the many investigations undertaken at the time was the search for porphyrins (precursors of

chlorophyll, hemoglobin, and other plant and animal pigments) in meteorites by G.W. Hodgson at NASA's Ames Research Center and B.L. Baker at
the University of Alberta. Hodgson and Baker tested several carbonaceous
chondrites, an enstatite, and the ordinary chondrites Bruderheim and Peace
River for the presence of porphyrins. They found porphyrins only in four of the
carbonaceous chondrites.[47]

Stable Isotopes

Because of the striking similarities between Peace River and Bruderheim, H.R.
Krouse and R.E. Folinsbee (University of Alberta) investigated whether their
$^{32}S/^{34}S$ ratios were also similar. They detected a difference in the $^{32}S/^{34}S$ ratio of
the two meteorites of less than 0.1%, which was judged to be the precision of
the analyses.[48]

Isotope Chronology

The noble gas (He, Ne, Ar, and Xe) contents and isotopic abundances were measured in Peace River and two other chondrites by M.W. Rowe, D.D. Bogard, and
O.K. Manuel (University of Arkansas).[49] The 10.6 g sample of Peace River they
received contained a large chondrule (~0.5 cm in diameter, 0.244 g in weight),
which was removed and analyzed separately from the three larger samples of
bulk material. The samples were heated until melted and the outgassed gases
analyzed in a mass spectrometer. Rowe et al. found the ^{129}Xe (radiogenic,
derived from the decay of ^{129}I) content in the Peace River chondrule to be
higher than in the three bulk samples. Excess ^{129}Xe in chondrules relative to
total meteorite had previously been reported by Merrihue for Bruderheim.[50]
However, they did not find the Peace River chondrule to be depleted in primordial (non-radiogenic) Xe relative to the total meteorite as had been the case in
Bruderheim. They also measured the Ne and Ar abundances, and the isotope
ratios of He, Ne, and Ar of Peace River, from which they were able to estimate its
cosmic-ray exposure age (Table 11.3).

P. Englert and W. Herr, in their study of exposure ages of chondrites include,
for purposes of comparison, the exposure ages of several meteorites studied by
other researchers.[51] The exposure age of Peace River is given as 30.7 Ma, in fair
agreement with Rowe et al.

P. McConville, S. Kelley, and G. Turner (University of Sheffield) developed
a laser probe for analysis of Ar from single 80 µm diameter laser pits that
release gas from 0.5 µg of sample. The gas is analyzed by mass spectrometer.
McConville and his colleagues found that, in the Peace River chondrite, Ar has
been retained by the shock melt vein and lost by the matrix; the vein is deficient

Table 11.3. Age estimates for the Peace River meteorite.

Method	Solidification age (Ma)	Cosmic-ray exposure age (Ma)	Gas retention age (Ma)	Reference
$^{40}K/^{40}Ar$			960	Taylor 1964a
Noble gases		32	960	Rowe et al. 1965
^{53}Mn		30.7		Englert and Herr 1978
Rb/Sr	4500			Shima and Honda 1967a
$^{40}Ar/^{39}Ar$		~40	450	McConville et al. 1988
K/Ar		35		Turner et al. 1990

in K, which they measured indirectly by the concentration of ^{39}Ar (fast neutrons produced by cosmic rays transmute ^{39}K present in the meteorite into ^{39}Ar).[52] The observation that Ar has been retained by the glass in the shock veins and lost by the matrix can be understood, they said, in terms of a range of plausible annealing histories. They suggested that Peace River was almost totally and uniformly outgassed about 450 Ma ago, possibly following a collision involving its parent asteroid. The future Peace River meteorite then cooled slowly (annealed) at low temperatures, in the range of 400–600 °C, at depths of greater than a few metres in the parent body, effectively resetting its K-Ar clock. A more recent fragmentation event about 40 Ma ago reduced Peace River to its pre-atmospheric size and exposed it to cosmic ray bombardment.[53] G. Turner and J.M. Saxton (University of Manchester) and M. Laurenzi (Consiglio Nazionale delle Ricerche, Pisa) made laser probe measurements of the fusion crust of Peace River; their data indicate that the fusion crust retained a memory of the meteorite's pre-entry cosmic ray exposure, which they give as 35 Ma.[54]

The levels of radionuclides, such as ^{10}Be and ^{26}Al, produced in meteorites by cosmic rays are of much interest to meteoriticists. Measurements of these radionuclides can provide information both on the history of the meteorites themselves and on the variations of the cosmic ray flux in space and time. Hisao Nagai (Nihon University) and his co-workers were interested in developing low-energy accelerator mass spectrometry (AMS) as an alternative to the usual high-voltage AMS; although less sensitive than the latter, lower energy AMS would permit the use of smaller accelerators for meteorite analysis.[55] They measured ^{10}Be and ^{26}Al in four iron meteorites and two chondrites, Peace River and Bruderheim; in these last two, metal fractions were separated from silicate fractions and both were analyzed. Peace River was found to contain in metal, ^{10}Be (5.08 ± 0.97 dpm/kg), ^{26}Al (5.96 ± 0.54 dpm/kg); in stone, ^{10}Be (21.2 ± 0.0.8 dpm/kg), ^{26}Al (55.9 ± 4.2 dpm/kg). Nagai et al. also determined these isotopic ratios in metal, $^{10}Be/^{9}Be$ ($1.81 \times 10^{-11} \pm 0.34 \times 10^{-11}$), $^{26}Al/^{27}Al$ ($3.04 \times 10^{-11} \pm 0.25 \times 10^{-11}$), $^{26}Al/^{10}Be$ (1.17 ± 0.25 dpm/dpm); in silicate, $^{10}Be/^{9}Be$ ($20.0 \times 10^{-11} \pm 0.7 \times 10^{-11}$), $^{26}Al/^{27}Al$ ($11.7 \times 10^{-11} \pm 0.9 \times 10^{-11}$), $^{26}Al/^{10}Be$ (2.63 ± 0.22 dpm/dpm). Later, by comparing the ^{10}Be and ^{26}Al content in the metal phase with that in the silicate phase, Nagai and his colleagues were able to estimate the irradiation hardness (or resistance, of the meteorite) or the spectral shape of the cosmic ray flux and the intensity of the irradiation, and to numerically relate these parameters to the size of the pre-atmospheric body and the sample depth in it.[56]

H.W. Taylor (University of Alberta) used his own K data and Ar data provided by R.E. Folinsbee to calculate the gas-retention age of Peace River.[57]

Summary

Peace River bears a striking similarity, chemically and mineralogically, to Bruderheim, another ordinary chondrite. Its chief claim to fame comes from the discovery—in the dark veins which pervade the body of the meteorite—of a new mineral, named *wadsleyite* in honour of the late Arthur David Wadsley, a renowned Australian crystallographer. The type specimen, #11550/PR4, is preserved in the University of Alberta Meteorite Collection.

12

The Skiff Meteorite

History

Sometime in 1966, Bill Nemeth, a farmer in southern Alberta near the town of Warner, found an interesting dark, heavy rock in his field in the NE 1/4, Section 31, Township 3, Range 4, west of the Fourth Meridian.[1,2] As a collector of petrified wood, Mr. Nemeth recognized this rock as being something out of the ordinary. He kept it for 12 years before deciding to send it to the geology department at the University of Alberta for identification. There it was identified as an ordinary chondrite weighing 3.54 kg (Figs. 12.1 and 12.2); after negotiating with R.E. Folinsbee over the telephone, Mr. Nemeth agreed in early February 1978 to sell it to the the University for $300.[3]

Science

The Skiff meteorite is a group H chondrite of petrologic type 4 and is, in some ways, similar to the Belly River meteorite (Chapter 6). The most common mineral present is olivine, of which 18% is of the fayalite (iron-rich) variety; the rest is forsterite (magnesium-rich) (Fig. 12.3). The second most common mineral is pyroxene, of which the dominant form is bronzite. As an H chondrite, Skiff contains more nickel-iron than a L chondrite, such as Peace River (Chapter 11); specks of this metal alloy can be seen glinting on a cut surface of the meteorite (Fig. 12.4).

The Skiff meteorite was cut into two parts: one piece, weighing 1.43 kg, was sent to the National Meteorite Collection in Ottawa; the other, weighing 2.029 kg, remains at the University of Alberta.[4]

Fig. 12.1. A cast of the Skiff meteorite.

Fig. 12.2. A cast of the Skiff meteorite made at the Provincial Museum of Alberta before sawing, showing the reverse side of the meteorite.

Fig. 12.3. Two chondrules (centre-left) and a hexagon-shaped crystal of fayalite (centre-right) in the Skiff meteorite seen in cross-polarized light. (25× magnification). Field of view 4.5 mm. For colour image see p. 289.

Fig. 12.4. Specks of nickel-iron alloy can be seen on the sawn face of the Skiff meteorite.

13

The Vilna Meteorite

History

The Vilna meteorite may hold the world record as the smallest meteorite find ever; that is to say, the smallest *total* mass of recovered meteoritic material from a witnessed meteorite fall (as opposed to the tiny grains and specks of dust often found accompanying much more massive meteorites, such as Bruderheim). As R.E. Folinsbee, L.A. Bayrock, G.L. Cumming, and D.G.W. Smith (1969), the authors of one scientific report, put it: "The writers are perplexed by a new manifestation of the law of diminishing returns in connection with the recovery of meteorites from falls in western Canada."[1] Balanced against this meagre result, however, is one of the most interesting stories of all Alberta meteorites.

> The bright bolide which produced the Vilna meteorite flashed across the skies of northern Alberta at about 18:55:30 M.S.T. on [Sunday,] February 5, 1967..., early enough in the night to be observed by several thousand people, from Fort McMurray in northern Alberta to Calgary and Vulcan in the south—a radius of 280 miles [450 km]. Initial reports came in very quickly over the telephone and were relayed by the RCAF, R.C.M.P., radio and television stations to Edmonton....These established the bolide as a detonating one, with detonations [audible sounds], centred about 100 miles [160 km] north-east of Edmonton. [Dr. L.A.] Bayrock at the Alberta Research Council and [Professor R.E.] Folinsbee at the [University of Alberta] Department of Geology asked over the radio and by press for long distance calls or letters from persons in the area of detonation who had actually seen the fireball—that is, who were outside or at uncurtained windows at the time of the event. Many calls were received on Monday and these indicated a sufficiently small prospective fall area to make a search worthwhile, though the ground was covered with 1 to 2 feet [30–60 cm] of snow at the time."[2]

It is apparent from the preceding paragraph that considerable importance is attached to whether or not a bolide is heard. This is because although it may be probable that pieces of the meteorite reached the earth surface, if they are small and few in number, they will be impossible to find. The size of the bolide comes into play here. Upon reaching lower altitudes of the atmosphere the shock waves produced by the bolides will be heard over a large area resembling "sonic booms" produced by airplanes. Often two or three or even more explosions are heard with a following rumbling noise.... The author [Bayrock] agrees with [the American meteorite collector] H.H. Nininger that unless sound effects are heard the search for the meteorite should be discontinued because of its small size and the consequent difficulty of finding it.[3]

To emphasize the point Bayrock noted that "in the period from 1957 to 1965, 12 bolides were investigated in Alberta, of which three produced intensive sound effects. Pieces of all three sonic bolide meteorites have been found while none of the others have been located."[4]

By Monday evening, February 6, a rough outline of the fall area had been established from telephoned information (Fig. 13.1), and three parties left Edmonton for a rendezvous at Lac La Biche on the north edge of the fall area, with instructions to do as much interviewing as practicable on the way.[5]

Doubly fortuitously, the all-sky camera at the Dominion Observatory's auroral observatory at Meanook, north of Edmonton, was making a 40 second exposure at the very time of the bolide and the fireball passed almost directly overhead (Fig. 13.2). It was hazy at Meanook, and stars did not show up in the photograph, although Jupiter (magnitude –2.5) did. The moving bolide first became visible on the photograph when its rapidly moving trace equalled the brightness of Jupiter; it appears that it took another 4–20 seconds after the camera shut off for the decelerating fragments to reach the "end point." The end point, or "retardation point," is where the fragments have lost all their interplanetary velocity and begin to fall to the ground with normal gravitational acceleration, as would any object dropped from that height.

On Tuesday evening at Lac La Biche, observational data culled from interviews of witnesses were collated, and a preliminary pattern of fall was established on the theory that the azimuth of the bolide, measured from the Meanook photograph, was 109°. The true azimuth of the bolide as it passed over Meanook was actually about 104°, and this resulted in an error of about 8 km in the predicted fall area.[6]

In the end, more than 300 reports were submitted (Fig 13.3). Mrs. C.L. Brickman, from Viking in a letter addressed simply to "Dr. L.A. Bayrock, Meteorite search director, Edmonton, Alberta," described how she saw a flash

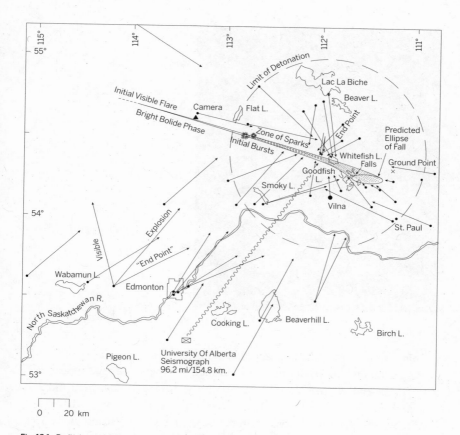

Fig. 13.1. Preliminary sighting data for the Vilna bolide received by telephone on 6 February 1967, allowed a rough outline of the fall area to be established.

through her north kitchen window, but it was gone by the time she and some dinner guests went outdoors to look.[7] Perhaps the best written description of the Vilna bolide came from Mrs. Dan Edmonton of Wabamun, in a letter dated 7 February 1967:

> My husband and I and another couple were headed for Stony Plain on Sunday evening and just before we had passed the weather station my husband turned around to see what was coming up behind us as the windshield had lit up. He drew our attention to the sky. I was seated in the back of the car and had to lean over another person to see out of the window. I saw a shape which I thought was the moon (immediately after I realized it was much smaller—it seemed about like a bread and butter

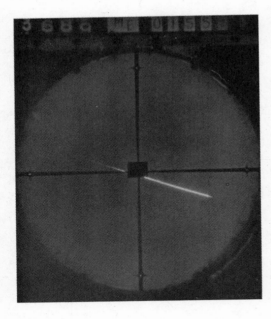

Fig. 13.2. The Vilna bolide as recorded by the Meanook all-sky camera.

plate) headed in an arc to the ground. I had just glanced at it a second or so when it momentarily disappeared and suddenly there was what appeared to be an explosion and sparks shot around profusely. It reappeared and continued downward. My husband said he saw two explosions with the fireball disappearing momentarily before each flash. He had also noticed a trail behind it. We were all so surprised we took note of the time—6:55 p.m. My husband and the other gentlemen felt it might have landed around Calahoo. What amazed us about the whole thing was the shower of sparks during the explosion when the meteor was still in the air. The car windows were closed and none reported hearing any sound.[8]

Mr. and Mrs. Edmonton were later interviewed by investigators at the point of observation, a hill with a commanding view to the north-east, and it was established that the fireball was first observed at [a bearing of] about 345° traveling with an observed apparent slope of 15° to the east. It exploded at 37° and an elevation of 11°, disappeared momentarily and reappeared as a shower of sparks that disappeared between 60° and 90° at an elevation of 4°. This is a very accurate observation....The endpoint was described as a pulsating zone of apparent retardation..."like someone firing retro-rockets to slow the fireball down."[9]

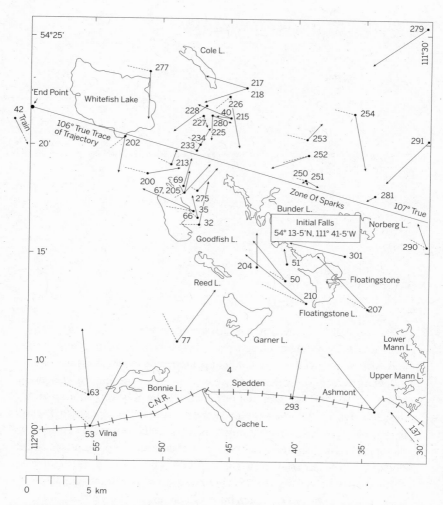

Fig. 13.3. Map showing sightings of the Vilna bolide (dashed lines: first sightings; solid lines: last "spark"). The numbers indicate observers (see text).

From a vantage point west of Wabamun Lake, Professor and Mrs. John Lauber saw the fireball to the northeast and wrote a very good description of it:

[on] February 5, 1967, 2 or 3 miles west of Entwistle, driving east on Highway 16 at 50 m.p.h. [80 km/h], I saw a light in the sky to the north-east, which reminded me of summer lightning. I quickly realized that it could hardly be lightning, and that we were too far away to see the lights of Edmonton, and

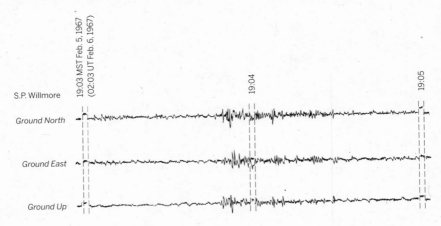

University of Alberta Seismograph 53-223N. 113-348W

19:03 MST Feb. 5, 1967
(02:03 UT Feb. 6, 1967)

19:04

19:05

S.P. Willmore

Ground North

Ground East

Ground Up

Fig. 13.4. Seismograms clearly record the arrival of the air blast from the Vilna bolide.

said to my wife, who was driving, "What's that?" We both saw a greenish glow, and as we watched, a bright green ball appeared about 45 degrees above the horizon and to our left. It appeared to be moving east, and following a steeply descending trajectory. I estimate the angle of descent at about 35°. We both thought that its light might be pulsating somewhat. It descended to about 20 degrees above the horizon and disappeared. Except for its colour, we might have thought that it was a large meteorite, very near to us. Its appearance was as follows: it was a large, apparently round, body, about the size of a streetlight seen from a block away; the colour was softer than the green of a traffic light or an airplane's landing lights; the light seemed to vary in intensity, but it did not have separate, individual lights—and was fuzzy at the edges. (We were looking through a rather dirty windshield, but the headlights of approaching cars, and the house lights that we passed did not appear as fuzzy as this object. We estimated that the ball itself was in view for not more than 30 seconds; the prior greenish glow for perhaps another 30 seconds....[10]

Arthur Page of the Meanook observatory was outside at the time the fireball passed over Meanook, at a point two miles (3.2 km) north of the all-sky camera, and he was able to provide Bayrock with very exact data on the azimuth and elevation of the explosion (burst or detonation); he estimated the luminosity

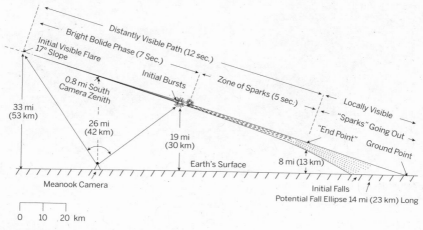

Profile Derived From Photographic And Sighting Data, Vilna Meteorite

Fig. 13.5. Profile of the path of the very slow Vilna bolide.

of the bolide as being that of the full moon (magnitude −12.6). The fireball was bright enough, even in its dying phases, to generate sharp, rapidly moving shadows: Wayne Erasmus, about four miles (6.4 km) east of Whitefish Lake, saw the shadows of poplar trees projected onto his bed in a darkened room with a south-facing window. The Vilna fireball was sufficiently bright to automatically turn off barnyard lights 15 miles (24 km) from the end point. James Wright, travelling north under a clear sky near Vulcan, 450 km from the end point, gave the magnitude as being about −4.5 (as bright as the planet Venus).

The stresses on the Vilna bolide during its fiery flight through the increasingly dense air of the upper atmosphere finally became too great, and it exploded at an altitude of about 30 km. Visual observations suggest that the fireball continued down to the end point at a height of 13 km (Fig. 13.5). The detonations accompanying the Vilna fireball were heard throughout the fall area at distances of 40 miles (64 km) or more from the end point. George Davy, 10 km north of the end point, heard four "booms" separated by intervals of 5–6 seconds. These four detonations were recorded by the University of Alberta's seismograph station southeast of Edmonton, 156–160 km from the end point (Fig. 13.4). The average atmospheric temperature at the time was 0 °C, giving a speed of sound of about 330 m/s; it took 470–485 seconds, or about 8 minutes, for the sonic boom to reach the seismograph. Several observers reported hearing hissing sounds accompanying the actual passage of the fireball.

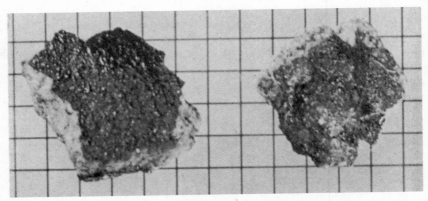

Fig. 13.6. The two recovered Vilna fragments, 94 mg (left) and 48 mg (right), showing the black fusion crust. Background squares are 1 × 1 mm.

Chuck Hires at Warspite described the sound as "like a sled or a toboggan going through snow."[11] However, there is no trace of this noise on the seismic records. The air blast from the Vilna bolide lasted about 40 seconds.

Fragments of the bolide—"sparks"—were seen only by people in the immediate Vilna fall area, though these were followed almost to ground level by enough independent observers (marked as numbers 225, 250, 251, 252, 253, 254, 281, 290, 291 in Fig. 13.3) to clearly establish their existence.[12] Mrs. Helen Calliou of the Métis settlement at Kikino was walking northwest along a dark road directly toward the bolide. Keenly observant, she first noticed it as a distant star growing in brightness but not moving much (she was almost directly along the line of trajectory). She also followed the last sparks down almost to ground level.[13]

The search for meteorites in the area of fall began by snowmobile on Wednesday, February 8, with particular attention being given to areas of drifting wind-blown snow: from experience gained in the Bruderheim search in 1960, it was suspected that small meteorite particles might be exposed by snow erosion. Sure enough, "on Feb. 9, Ted Reimchen, a student in geology at the University of Alberta, recovered from the snow surface on a small lake a small flat piece of meteorite with a black fusion crust on one side, weighing 48 milligrams [Fig 13.6]. A second slightly larger fragment (94 mg) was recovered a few feet away, partly buried in the snow, by Jack Grant of the Meanook observatory on Feb. 11....By this time it was snowing heavily and all chances of further recovery from the snow surface disappeared."[14] The fall of the Vilna meteorite was reported to the *Meteoritical Bulletin*.[15]

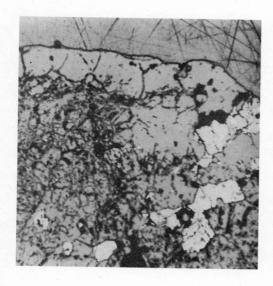

Fig. 13.7. Photomicrograph of a polished surface of Vilna #2, seen in reflected light, showing the fusion crust and a number of troilite grains (light) in a silicate matrix. The fusion crust, near the top, is approximately 100 μm thick. The scratched area at the top is an artifact of the polishing process

Later in the spring, the search shifted to the area north of Norberg Lake, a sparsely populated region with little cultivated land. The search concentrated on the lake surfaces until melting of the ice in May made this impossible. No new recoveries were made. Late spring snow persisted and made ground searching very difficult. More ground searching was done during the summer and autumn without success.

Science

Folinsbee and his colleagues polished the larger of the two fragments and studied it optically and by electron microprobe. They did a *modal analysis* by counting 1000 points on the polished surface (16 mm²) using a Zeiss integrating ocular (an automatic point counter) attached to a Leitz metallographic microscope. They identified the points as 85.6% silicates and phosphates, 10.2% kamacite and plessite, 3.9% troilite, and 0.4% chromite. Thus, Vilna proved to be an ordinary chondrite, quite similar to the Bruderheim and Peace River meteorites.[16] The slightly vesicular (bubbly and pitted) black fusion crust shows quite clearly in the polished thin section (Fig. 13.7).

The results of the electron microprobe were not reported until four years later, by D.G.W. Smith, R.E. Folinsbee, and M. Hall-Beyer.[17] Their quantitative analysis of the composition of the minerals present in Vilna—olivine, orthopyroxene, clinopyroxene, plagioclase (or possibly maskelynite), chromite, whitlockite (=merrillite), apatite, troilite, kamacite, and taenite—confirmed the earlier identification of the meteorite as a chondrite. Their results also led them to identify Vilna more specifically as an L chondrite of petrologic type 5 or 6.

14

The Vulcan Meteorite

History

The Vulcan meteorite was found by Mr. Naaman Budd in April 1962, in about 30 cm of soil on his farm 12 km northeast of the town of Vulcan (Fig. 14.1).[1] Mr. Budd brought the meteorite in to the Department of Geology, University of Alberta at Calgary, in March 1964.[2]

The stone was examined by Dr. T.A. Oliver, chairman of the department. He confirmed that it was a meteorite, weighed it (19.025 kg) and noted that the fusion crust was mostly intact, suggesting that it was a fairly recent fall. On the other hand, considerable oxidation (i.e., rusting) of the meteorite's interior was also reported, indicating that it may have lain in the ground long enough for air and water to have seeped through cracks in the rock.[3] The meteorite was cut into three approximately equal pieces (Fig. 14.2): one of these was retained in Calgary; one, probably the middle section, was shipped to the University of Alberta in Edmonton; and the third was sent to the Geological Survey of Canada's National Meteorite Collection in Ottawa. After this sectioning, the Vulcan meteorite was further divided and distributed: the Natural History Museum in London obtained an 880 g specimen; The University of California, Los Angeles, has a 165 g sample; and Robert A. Haag, a private collector, and several museums and institutions have lesser bits and pieces. Notwithstanding this dissemination, the Vulcan meteorite has been comparatively poorly studied.

Science

The Vulcan meteorite is a stony meteorite, more specifically an ordinary chondrite of class H6. This means that it contains a high proportion (15–20%) of metallic iron. With so much iron tied up as metal, less is available to form minerals such as fayalite, an iron-rich variety of olivine. Mason (1967) gave Vulcan's olivine composition as Fa_{20} (meaning that it consists of 20% fayalite) and classified the meteorite as an olivine-bronzite chondrite.[4]

Measurements of the levels of radioactivity from ^{26}Al produced in meteorites by exposure to cosmic rays have been widely used to obtain a variety of information both on the history of the meteorites themselves and on the

Fig. 14.1. Mr. Naaman Budd (left), Mrs. Vernice Budd, and their son Warden Budd (right) holding the Vulcan meteorite.

variations of the cosmic ray flux in space and time.[5] This isotope of aluminum (ordinary aluminum consists entirely of the non-radioactive ^{27}Al isotope) has a half-life of 0.74 Ma and it emits gamma rays as it disintegrates, or decays, into a stable isotope of magnesium, ^{26}Mg. Ian Cameron and Zafer Top (University of New Brunswick) measured the level of ^{26}Al in 50 stone meteorites including Vulcan, using a gamma-ray spectrometer.[6] They used a sample from a 386 g slice of Vulcan provided by the Geological Survey of Canada from the National Meteorite Collection of Canada. The sample was mounted on the front face of one of the spectrometer's two scintillators (these contain a scintillation liquid that gives off tiny flashes of light, one per gamma ray, that are counted by photosensors); the other was used to detect any background (non-meteorite) gamma radiation. Cameron and Top measured an activity of 53 ± 2 dpm/kg (disintegrations per minute per kilogram of sample). The interpretation of ^{26}Al activity in terms of meteorite orbits is, as they say, "subject to considerable error from a variety of factors."[7] Nevertheless, they were able to postulate an orbit for the Vulcan meteorite with an aphelion in the range of 1.95–2.00 astronomical units; in other words an orbit that, at its farthest from the Sun, lies just outside the orbit of Mars on the inner edge of the main asteroid belt.

Fig. 14.2. Rough unpolished sawed slab of the Vulcan meteorite.

The scientific investigation of a meteorite can involve sawing, chipping, crushing, grinding, polishing, and dissolving, which results in the destruction of some or all of the specimen. For this reason, meteoriticists often create a permanent record of the original specimen by making a cast or replica of it. Such copies are sometimes even painted and weighted to match the original meteorite in appearance and feel. University of Alberta scientists, wanting to make a realisitic copy of their slab of Vulcan meteorite, developed the following recipe (Fig. 14.3).[8]

Mock Meteorite

Ingredients:

Mold of meteorite, previously prepared

643.0g	Lead, powdered
571.6g	Iron filings
214.3g	Plaster of Paris
71.4g	Sanidine (potassium feldspar), crushed

Water, as needed

Instructions:
In a bowl, combine dry ingredients and add just enough water to produce a pourable mixture.
Pour mixture into a mold and allow to dry. Remove replica from mold when hard.

Fig. 14.3. Replica meteorite recipe.

15

The Abee Meteorite

History

Monday, 9 June 1952, was a pleasant late-spring day in north-central Alberta, although a few clouds that presaged rain later in the week drifted in from the west. At this latitude (54°N), there is continuous twilight during June nights, and many people were still outside at 11:00 p.m. enjoying the warm, bright evening. Unbeknownst to them, far overhead a substantial chunk of rock—a fragment of an asteroid—was just encountering the thin air of the upper atmosphere, its ancient orbit through space having intersected that of the Earth. As the rock's tremendous velocity, perhaps 20–40 km/s, was slowed by friction with the increasingly dense atmosphere, the air in front of the rock was violently compressed and heated into incandescence. Meanwhile, the outer surface of the rock was being ablated or melted and burned off; perhaps two-thirds of the the original mass (estimated at about 300 kg) was lost due to atmospheric ablation.[1] Still 75 km or more above the surface of the Earth, the rock must have been a bright meteor, "but the fact that it approached from the direction of the persistent twilight, combined with the broken clouds, indicates that it was not noticed until it became extremely bright at lower heights."[2]

At 11:05 p.m., the now dazzling fireball burst into view from the twilight sky. One Edmonton witness said that it

appeared first above a thin layer of clouds. It swung in a fairly low arc through the clouds, then split into two portions before reaching the earth.... Mrs. Fritz Smith, 9113 Saskatchewan Dr., said the object, about the size of a basketball, seemed to move down the North Saskatchewan river valley trailing small fragments. The [Edmonton] Journal and the Dominion Weather Office at the Municipal Airport both received phone calls from citizens who said they saw a red "ball of fire" with a blue streak behind it, passing over the city. Residents of Newbrook [about 80 km north of Edmonton] were startled by the flash of light and an explosion, followed by a low rumbling sound. A group of school children, returning home from a wiener roast, were frightened when they saw "flashing lights" and heard an explosion in the sky.[3]

Ironically, the well-equipped Newbrook Observatory (see Chapter 16), only 13 km north of Abee where the meteorite fell, was shut down: "Because of the bright nights, meteor photography is not practical near the summer solstice and [Arthur Griffin, the man in charge of the observatory] was indoors. Although [Griffin] did not witness the fireball, he heard the associated sonic booms, a sound similar to distant artillery fire that persisted for about a minute."[4] Other sightings were reported from Athabasca and Thorhild. At Spedden, about 185 km northeast of Edmonton, "a witness in a drive-in theatre reported that the light from the fireball blanked out the picture."[5]

In their report on the fall of the Abee meteorite, Griffin, Peter Millman, and Ian Halliday described how the meteorite was recovered (Fig. 15.1):

When Harry Buryn spotted a deep hole in his freshly seeded wheat field on Saturday, June 14, he mentioned the unusually large hole to his neighbour, a school teacher. She thought of the fireball and informed the first author [Griffin] late on June 15. When he arrived on the scene with a camera the following morning, a group of students had already excavated the meteorite. Small students had climbed into the hole and secured ropes around the stone to haul it to the surface. As previously reported by Dawson et al. (1960) the hole was about 0.7 m in diameter and 1.5 m in depth, an indication of the penetrating ability of a sizable meteorite in cultivated soil. The hole had a slope of about 25° to the vertical, with the bottom displaced to the southeast, indicating the direction of arrival was approximately from northwest.[6]

The location of the fall was determined to be 54°12'55"N, 113°00'23"W.

The meteorite was first taken to the hardware store in Newbrook where its weight was established as 107 kg, and it was prominently displayed for a few weeks. Later, it was moved to the grain elevator in the hamlet of Abee until arrangements were completed for its purchase by the Geological Survey of Canada [GSC]....[7,8]

Eugene Poitevin, Chief of the Mineral Division and Collection Curator of the Geological Survey of Canada determined it to be an enstatite chondrite—a rare type of stone meteorite. Poitevin offered $1,062 or $10 per kg to Buryn and Abee, then the third largest meteorite to be found in Canada, was acquired by the GSC for the Canadian National Meteorite Collection in Ottawa.[9]

Fig. 15.1. Mr. and Mrs. Harry Buryn with the Abee meteorite just after its recovery, 16 June 1952.

Science

As explained in the Introduction, chondrites are classified by their chemical and mineralogical composition. The most common or ordinary chondrites are classified as H, L, and LL, depending upon their iron (Fe) content and. The rare *C*, or carbonaceous, chondrites are carbon rich, whereas the equally rare and peculiar *E*, or enstatite, chondrites consist largely of the mineral *enstatite*, which is a variety of pyroxene with relatively little or no Fe. Instead, Fe is present as unoxidized metal and iron sulphide in these meteorites. They also often contain very unusual minerals like *oldhamite* and *sinoite*, which are never seen in Earth rocks, being stable only in environments that are extremely depleted in oxygen. Enstatite chondrites have many analogies in bulk composition to the ordinary chondrites; for example, they can be classified into EH (high Fe content) and EL (low Fe) groups. Ordinary chondrites and, more recently, enstatite chondrites have also been classified on a scale ranging from 3 to 6 according to their petrologic type; that is, on the extent to which they have been modified by heat, or *metamorphosed* prior to arriving on Earth. As this classification scheme has evolved, Abee has been variously classified as an E4, EH, and EH4 chondrite.[10]

Mineralogy and Textures

After its purchase by the GSC, Abee was shipped to Ottawa, where several plaster replicas were created for future reference and display purposes; in 1960, the meteorite was subjected to an intensive study by K.R. Dawson and J.A. Maxwell of the GSC and D.E. Parsons of the Mines Branch, Department of Mines and Technical Surveys.[11] They described it as an angular block measuring 30.5 × 40.6 × 47.0 cm, with a volume of 28,400 cm³, a mass of 107 kg, and a specific gravity of 3.5. "The general shape is characterized by planar surfaces and angular corners which have been rounded by its passage through the atmosphere. Locally, the surface exhibits a variety of irregularities which include a triangular-shaped scar, areas of spalled crust, regmaglypts, flutings and a pebbled surface texture [Figs 15.2 and 15.3]."[12] Dawson and his colleagues decided to first saw the meteorite in half at its equator and one of these halves was polished to reveal the internal structure. They then made a second cut parallel to the first, producing a triangular plate approximately 2.5 cm thick that was then cut into variously sized fragments for study by several different methods. They described the Abee meteorite as a "'polymict' chondritic breccia' containing angular fragments, or *clasts*, of black chondrite, ranging in size up to 8 × 6 cm, in a fine-grained matrix of similar material.[13] The fragments include smaller pieces of metallic iron and iron-silicate mixtures; many fragments also contain black, spherical chondrules which rarely exceed 1 mm in diameter. The meteorite contains no fractures or veins, and it breaks across rather than around the chondrules."

Dawson et al. identified the constituent minerals in powdered samples of Abee using an X-ray diffraction technique, including abundant enstatite and the nickel-iron alloys kamacite and taenite; minor troilite; and accessory oldhamite, plagioclase, and plessite (a mixture of kamacite and taenite). They also found alabandite, quartz, and α-*cristobalite*, "which have not been previously reported in stony meteorites." They suspected but were unable to prove the presence of *cohenite*. About this time, Harold Urey and Toshiko Mayeda (University of Chicago) provided a detailed description of the metallic particles in Abee and several other chondrites.[14] They examined polished and acid-etched sections of the meteorites microscopically using reflected light. In all but the very smallest particles, they were able to identify the different metallic minerals: kamacite has a white luster; taenite has a very slight yellow; troilite is distinctly yellow and commonly cracked; plessite varies greatly, "from a slightly cloudy appearance to a coarsely crystalline character."

The Abee meteorite is by far the largest enstatite chondrite known; because of this, it is among the most widely distributed and extensively studied

Fig. 15.2. Side view of the Abee meteorite showing its form, areas from which the fusion crust has spalled, and regmaglypts. Maximum diameter is 40 cm.

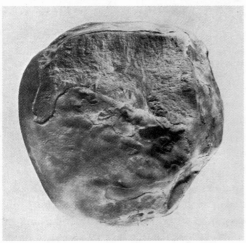

Fig. 15.3. Vertical view of the Abee meteorite showing its form, the scar area, regmaglypts, flow lines, and spallation areas on the fusion crust. Maximum diameter is 40 cm.

meteorites in the world. Some 20 years after the Dawson group's study, Kurt Marti (University of California, San Diego) established the Abee Consortium, a group of 13 researchers from nine different institutions in three countries

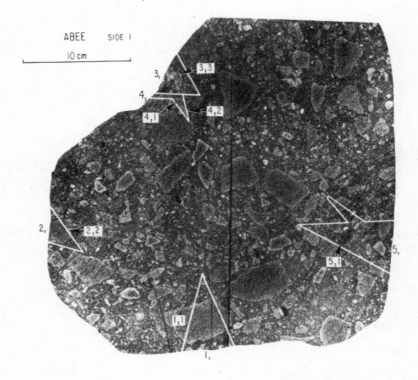

Fig. 15.4 One side of the slab from the Abee meteorite used in the Abee Consortium study. The triangular sample cuts (1–5,) and the clasts studied (1,1; 2,1; 2,2; etc.) are indicated. Slab is 32 cm wide (left to right).

(USA, Canada, and India), to examine the Abee meteorite in detail.[15] They reported their results in seven papers published concurrently in 1983 in a single issue of *Earth and Planetary Science Letters*. Alfred A. Levinson (University of Calgary) generously provided the consortium with a large slab through the central region of Abee. Five triangular pieces were carefully cut from the periphery of the slab and distributed to the investigator teams; these, plus individual clasts and other special samples, were carefully documented and mapped (Figs. 15.4 and 15.5).

Alan E. Rubin and Klaus Keil (University of New Mexico) examined the mineralogy and petrology of the Abee meteorite for the consortium.[16] They prepared several polished thin sections of clasts, matrix, chondrules, and dark inclusions from the samples cut out of the slab. No water was used in the preparation of the sections in order to avoid alteration of water-soluble minerals;

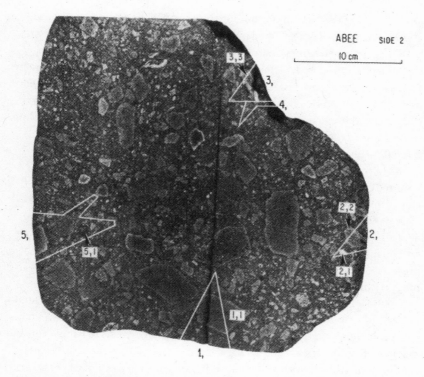

Fig. 15.5 The reverse side of the Abee meteorite slab depicted in Fig. 15.4. By careful comparison of both photographs, it is possible to visualize how individual clasts change in size and shape from one side of the slab to the other. The triangular sample cuts and clasts are again indicated. Slab is 32 cm wide (left to right).

samples were cut on a wire saw using isopropyl alcohol, cleaned with acetone, and ground and polished using kerosene.[17] Rubin and Keil studied the sections microscopically in transmitted and reflected light, using a point counter for *modal analyses*; mineral analyses were made with an electron microprobe. They reported that mineral compositions are "remarkably uniform throughout the clasts and matrix, but the dark inclusions are more heterogeneous." They described the irregularly shaped clasts, which make up about 50% of the meteorite, as measuring up to 6 cm in maximum extent. Chondrules, 0.2–2.0 mm in diameter and composed mostly of pyroxene but also containing enstatite, occur throughout the meteorite. The so-called dark inclusions range up to 1.3 cm in maximum dimension; contain much less silica and metallic Fe and nickel (Ni) but more total sulphides (e.g., oldhamite) than the clasts and matrix; and are probably darker coloured on account of their finer grain size. Abee appears to be unique among the enstatite chondrites for its large abundance of *niningerite*, named after the American meteorite collector and authority Harvey H. Nininger; Rubin and Keil reported values of up to 15% by weight in clasts and dark inclusions. They also reported up to 9% oldhamite in dark inclusions. Many clasts, some chondrules, and a few dark inclusions have metallic Fe- and Ni-rich margins or rims. In thin section, the boundaries of the clast and dark-inclusion matrices are not sharp. Many areas of Abee, including the borders of chondrules, are recrystallized and chondrules are sometimes difficult to recognize.

Geophysicists N. Sugiura and D.W. Strangway (University of Toronto) reported on the thermal and magnetic history of the Abee meteorite for the consortium.[18] When the parent bodies of meteorites first coalesce from the solar nebula, the magnetic particles in their molten rock align themselves according to the prevailing magnetic field, much like iron filings surrounding a small magnet. When the rock solidifies, this alignment is "frozen" into place, preserving a record of the direction of the Sun's magnetic field. This frozen-in magnetism is referred to as *paleomagnetism*, or *natural remanent magnetization* (NRM). The NRM is erased and reset when (and if) a meteorite is reheated either during its agglomeration or subsequently. Because of the large size of its clasts and the strength of its NRM, Abee is well suited to tests of its thermal and magnetic history. Sugiura and Strangway reported that the magnetic orientations of several clasts are significantly different from each other, whereas that of the matrix is uniform, suggesting that the clasts were magnetized before agglomeration and that the meteorite was not reheated to temperatures much above 100 °C. They also reported that samples taken from near the meteorite's fusion crust show evidence of the NRM being disturbed by heating during atmospheric entry. In a previous paper, Sugiura and Stangway reported that the NRM was carried in Abee by the mineral cohenite.[19]

Derek W. Sears, Gregory W. Kallemeyn, and John T. Wasson (University of California, Los Angeles), in the last of the Abee Consortium papers, succinctly summed up the peculiar fascination Abee holds for researchers:

> The Abee meteorite is of special interest because it is the largest (107 kg) of a rare and unusual clan of meteorites, the enstatite chondrites. Only 2% of observed falls are members of this clan and only 23 are known [as of 1983]. To add to its particular interest, Abee is brecciated, containing clasts of up to several centimeters in dimension (some metal-rich), and some 5-mm-sized dark inclusions. [20]

Sears and his co-workers go on to note the extraordinary degree of reduction in enstatite meteorites (see Table 15.1): silicon (Si) is present in the metal; elements like manganese (Mn), titanium (Ti), chromium (Cr), and calcium (Ca) that would normally be found as silicates, are instead present as sulphides; and the major silicate mineral (enstatite) is almost Fe-free, the iron being almost totally metallic (unoxidized).[21] This is in sharp contrast to the other classes of chondrites where 50%–100% of the Fe is oxidized. Sears et al. measured the concentrations of 25 major, minor and trace elements in four Abee clasts, a metal-rich inclusion and two dark metal-poor inclusions, using neutron activation analysis techniques. They reported that the Abee clasts differ in composition, displaying twofold enrichments or depletions in some elements. They suggested that this heterogeneity originated right at the beginning during the agglomeration-accretion process. The dark inclusions are more enigmatic: they are compositionally very different from the meteorite as a whole, being rich in both sulphides and rare earth elements; they seem to occur only in the matrix of the breccia; and they are not chondrules. Also, although the two dark inclusions studied are similar to each other, they are very different from any known class of meteorite or any reasonable interpolation between known meteorite classes.

A third exhaustive treatment of Abee's mineralogy was published in 1997 by Alan Rubin (University of California, Los Angeles) and Edward Scott (University of Hawaii, Honolulu).[22] Rubin and Scott presented a revised scenario for the geological history of Abee that emphasizes the role of impact events. Shock features went unrecognized in enstatite chondrites (e.g., Dawson et al. reported an absence of fractures or veins in Abee, while the Abee Consortium made no mention of shock features) until Rubin, Scott, and K. Keil extended the shock-metamorphism classification scheme of Stöffler et al. (see Chapter 6) to account for shock effects in the mineral orthopyroxene.[23] Rubin and Scott studied polished thin sections of Abee and some 24 other EH chondrites microscopically in transmitted and reflected light, identifying signs of thermal metamorphism, brecciation, and shock metamorphism. In addition to the aforementioned orthopyroxene, Rubin and Scott found clear evidence of shock, ranging from S2 (very weakly shocked) to S4 (moderately shocked) in Stöffler's scheme, in enstatite grains. They noted the extensive darkening of silicate minerals, a feature widely attributed to shock heating; the evidence of rapid cooling, which is characteristic of impact heating or quenching; and the presence of shock-produced diamonds. They suggested that Abee experienced two episodes of impact melting with an intervening period of brecciation.[24] The precursor rock of Abee was an EH3 or EH4 chondrite, which was partially melted during the first impact event, producing a relatively homogeneous, non-friable rock with a texture similar to that of the present clast interiors. Meteoroid bombardment fractured this rock, creating fragments of various sizes. During this period, the dark inclusions were introduced to the Abee *regolith* (the loose material covering the surface of the parent body) from a region of different composition, or, possibly, from a different parent body altogether. A second major melting episode flooded the Abee region with molten enstatite chondrite material, which digested (melted) small clasts—explaining why most Abee clasts are quite large (~0.6–6 cm)—then cooled rapidly, forming the fine-grained ground mass or matrix of Abee. A still later impact must have broken up this Abee material, flinging chunks of debris into space; one such piece eventually encountered the Earth and fell on Alberta.

As previously discussed in Chapter 2, the beautiful and distinctive pattern seen on the polished and etched surface of an iron meteorite is the result of the originally hot meteorite having cooled very slowly at the rate of perhaps a few degrees per million years. However, enstatite chondrites, show evidence of having cooled from an elevated temperature in an extremely short time. J.M. Herndon and M.L. Rudee (University of California, San Diego) conducted thermal experiments on laboratory-prepared iron alloys and samples of Abee. They then microscopically examined and compared polished and etched sections of alloy metal with Abee material. They concluded that the metal phase of Abee cooled from above 700 °C to room temperature in less than 10 hours. Herndon and Rudee suggest Abee must have been a very small body to have cooled so rapidly.[25]

The examinations of Abee by Dawson et al., the Abee Consortium, and Rubin et al., as extensive and detailed as they were, reveal the pitfalls of basing analyses of a chondrite on samples that, although in themselves quite substantial, represent a small portion of the meteorite as a whole. Russell Kempton, the Director of New England Meteoritical Services (Mendon, Massachusetts), reported that:

> To date, all studies have been conducted on small pieces originally sectioned from one end of the meteorite around 1960. Recently [1996], however, another [20 kg] section—the oriented end piece—has been made available for study. The surface dimensions in both axes are 38 cm × 34 cm offering approximately 1,020 cm² for a rare macro view of the interior that differs in some ways from other reported sections.

The most obvious structure in this end piece is a huge clast 22 cm × 14 cm covering approximately one-third of the surface. This clast and three others in this specimen continue through the meteorite to the exterior surface. the fusion crust is eroded around these clasts with the result being a clastic outline on the crusted exterior. The breccia displays a distribution of irregularly shaped clasts in a 50/50% volume representation of 1 mm to 4 cm clasts and eight larger 6 cm to 22 cm clasts. This is different from the irregularly shaped 1 mm to 6 cm size clasts [previously] reported. These two clastic groups, found in this end piece 21 cm from sections previously studied, indicate that Abee may be a transitional rock displaying the beginning of a bimodal distribution of fragments.

Chondrules are minimal within the clasts supporting the observations of Rubin and Keil (1983). Two clasts, 5.5 cm × 4 cm and 5 cm × 3.5 cm, differ from previous reports. Both have a higher abundance of chondrules in comparison to the other clasts and may be similar to those reported in Adhi-Kot [Pakistan; another EH4 chondrite]. All clasts are outlined with coarse-grained, rind-like metal rims.[26]

Kempton reported that Abee has been reclassified from an EH4 to an EH4 *impact-melt breccia*,[27] the only one known, and comments: "Everything about this meteorite is interesting." He then hypothesized that Abee may have originated from the primordial surface of Mercury, presenting a scenario in which conditions in the early solar system near the orbit of Mercury could have both produced chondritic material and resulted in its ejection from the inner solar system.[28]

The large section of Abee, referred to by Russ Kempton (above), was made available for dissemination and research in what may have been the most inspired "deal" in the history of Canadian meteorite collections.[29] Richard Herd, Curator of the Canadian National Meteorite Collection in Ottawa, was able to trade the Abee piece in order to acquire the Schmidt collection of over 900 specimens, representing about 450 different meteorites, from the estate of the late Terry Schmidt of Colorado Springs. In exchange for receiving the Abee specimen, Kempton agreed to reimburse the Schmidt estate for the collection. The new specimens, which included two samples of Martian meteorites, nearly doubled the Canadian National Meteorite Collection's holdings.

In 1968, Klaus Keil (University of New Mexico) reported the results of a new mineralogical and chemical study of 15 enstatite meteorites, including Abee.[30] Keil first measured the mineralogical composition of Abee by point-counting a small area (6.4 cm^2) of several polished sections of the meteorite under the microscope. He found that Abee consisted of about 24% nickel-iron alloy, or kamacite; 6% troilite (FeS); 11% niningerite, and some 58% enstatite. Keil then determined the chemical composition of the minerals by electron microprobe analysis. He found that the enstatite consisted of about 99% MgO and SiO_2, and less than 1% oxidized iron (FeO). One notable feature of enstatite from enstatite chondrites is its intense red and/or blue luminescence under electron bombardment. Keil commented on this phenomenon, which was subsequently developed by other workers into a new analytical technique called *cathodoluminescence* (CL).

Yanhong Zhang, Shaoxiong Huang, and their colleagues at the University of Arkansas examined the CL and *thermoluminescence* (TL; see Chapter 10) properties of some 36 enstatite chondrites, with a view towards understanding the thermal history of these meteorites.[31] They report that the enstatite in metamorphosed EH chondrites like Abee displays predominantly blue CL, whereas the enstatite in metamorphosed EL chondrites displays a distinctive magenta CL. These differences and the differing TL properties of the two classes of E chondrites found by Zhang et al. confirm earlier suggestions that metamorphosed EH chondrites underwent relatively rapid cooling, whereas the metamorphosed EL chondrites cooled more slowly.

R.M. Frazier and W.V. Boynton (University of Arizona) commented on Abee's many unique and intriguing properties:

> It is atypical in its high niningerite abundance, and it is also the only chondrite escaping the Re/Ir fractionation (Rambaldi and Cendales, 1980). It is also the only brecciated enstatite chondrite (Keil, 1968), containing 50% clasts 1–6 mm in diameter (Rubin and Keil, 1983); thus it may offer samples from different regions of the parent body(ies).
>
> Furthermore, Abee's mineralogy is representative of all enstatite chondrites: for example, it contains a low amount of MnO in the pyroxene.... It contains relatively low amounts of dissolved silicon in its kamacite and has highly recrystallized silica, all of which are characteristic of E5 chondrites (Rubin and Keil, 1983), even though Abee is an E4 for other reasons, such as its high volatile content.[32]

At the University of Chicago, C.A. Leitch and J.V. Smith made a detailed petrographic and mineralogical of four EH chondrites (Abee, Adhi-Kot, Kota-Kota, and Indarch) using optical microscopy, cathodoluminescence, and electron microprobe techniques (Fig. 15.6).[33] They described Abee as containing numerous clasts with metal-rich rims set in a fine-grained matrix; unlike other investigators, they found no chondrules.[34] Leitch and Smith found no olivine grains; but they did find small grains, with the composition $Mg_2Si_5O_{12}$, randomly scattered in the matrix of Abee that "require further study to check whether they are a new mineral." They suggested that their discovery of both blue and red cathodoluminescent crystals of enstatite enclosed in metal clasts "is sufficient evidence against a simple model" of origin of the EH chondrites by condensation from the solar nebula and that any new model must include mechanical aggregation of diverse materials formed in different regions of the solar nebula.

In the course of preparing the description of niningerite (named in honour of Harvey H. Nininger by Klaus Keil) for a mineralogical encyclopedia, Masaaki Shimizu (Toyama University, Japan) and Joseph A. Mandarino (Royal Ontario Museum) realized that this new mineral was really two minerals: niningerite (Mg,Fe)S, and an unnamed iron-dominant analogue (Fe,Mg)S.[35] A proposal to define and name the latter *keilite*, after Keil, was submitted to the Commission on New Minerals and Mineral Names of the International Mineralogical Association. Both mineral and name were approved (IMA 2001–053), and the *type locality* was designated as the site where the Abee meteorite was found. This is the second similar honor accorded an Alberta meteorite: the Peace River meteorite is the *type specimen* of another new mineral, wadsleyite (see Chapter 11).

In 1968, at the annual meeting of the Meteoritical Society in Cambridge, Massachusetts, J.A.V. Douglas and A.G. Plant (Department of Energy, Mines and Resources, Ottawa) made the first report of an *amphibole*, a silicate mineral, in an enstatite chondrite.[36] They had found clusters of 3.5 mm long crystals of *richterite* in the Abee meteorite. Edward Olsen (Field Museum of Natural History, Chicago), J. Stephen Huebner (U.S. Geological Survey, Washington, DC), and Douglas and Plant later presented data from their electron microprobe analysis of the Abee richterite.[37] Olsen et al. explained that "[meteoritic] amphiboles are an excellent potential water barometer or indicator of the presence or absence of significant H_2O in the environment in which the amphibole formed." In the case of Abee, however, they could "find no compelling evidence for the presence of structural water in the amphibole, and thus in the environment in which the amphibole last equilibrated."

A French research team led by Audouin Dollfus (Observatoire de Paris, Meudon, France) measured the reflectivity of a large variety of terrestrial, lunar, and meteoritic samples (including Abee) in the spectral range from 230 to 650 nm, i.e., from ultraviolet to red, to facilitate the future study of the surfaces of planets and asteroids by spacecraft and space-based telescopes.[38] About 6% of

Fig. 15.6 A polished thin section of the Abee meteorite viewed in reflected light. The light areas are highly reflective nickel-iron alloy. Note the round relict chondrule near bottom right. (40× magnification). For colour image see p. 290.

the meteorites that fall to Earth have been identified as possibly having come from the asteroid Vesta, because they share the latter's unique spectral signature indicating the presence of pyroxene. A comparison of Abee's reflectance spectrum to the spectra of different asteroids suggests that it may have come from the the asteroid Psyche.[39]

Minor and Trace Elements

Chondrites are of particular interest to scientists because they are the oldest, most primitive, least altered material we have available to study the early history of our solar system. Abundances of non-volatile elements in these meteorites generally agree well with those abundances that can be accurately measured in the Sun. H. Von Michaelis, L.H. Ahrens, and J.P. Willis (University of Capetown) measured the concentration of several elements in Abee and some 50 other stony meteorites (Table 15.1).[40,41] They also presented a critical assessment of the quality (precision and accuracy) of their analytical data.[42] R.A. Schmitt, G.G. Goles, R.H. Smith (Gulf General Atomic, San Diego), and T.W. Osborn (University of California, San Diego) measured the abundances of eight major and minor elements in 103 chondrites and 17 achondrites, using instrumental neutron activation analysis (INAA).[43] They acknowledged that: "The inhomogeneous distribution of silicate, metal and sulphide phases [in chondrites] may introduce severe sampling uncertainties." About the same time, Brian Mason (U.S. National Museum) reported determinations by several workers of varying rhodium (Rh), palladium (Pd), and platinum (Pt) abundances in Abee.[44]

M. Ikramuddin, C.M. Binz, and M.E. Lipschutz (Purdue University) determined the levels of 10 trace elements retained in samples of Abee heated at 400–1000 °C for one week in a low-pressure (~10^{-5} atm) hydrogen (H_2) environment.[45] They were interested in establishing the *volatility* and *mobility* of these elements—Co, silver (Ag), gallium (Ga), selenium (Se), caesium (Cs), tellurium (Te), zinc (Zn), bismuth (Bi), thallium (Tl), and indium (In)—at the temperatures and pressures believed to exist during the formation of the Abee parent body (Table 16.2). Ikramuddin, Lipschutz, and E.K. Gibson subsequently repeated their experiment, with the exception that the apparatus was purged with neon (Ne) gas rather than hydrogen.[46] They found the substitution of Ne for H_2 did not change the retentivity of the elements studied, except for Zn (in which the retentivity decreased significantly), and possibly In. A few years later, another Purdue University team (Swarajranjan Biswas, Thomas Walsh, Gerhard Bart, and Michael E. Lipschutz) repeated these elemental determinations at higher temperatures (1000–1400 °C).[47]

J.C. Laul, R. Ganapathy, E. Anders, and J.W. Morgan (University of Chicago) measured nine volatile elements in a suite of some 40 H, LL, and E chondrites, using NAA (Table 16.2a).[48] They were especially interested in the most volatile metals (Tl, Bi, and In), which can be used as "cosmothermometers" to indicate the accretion temperature of a meteorite. In the case of Abee, Laul et al. calculated highly concordant accretion temperatures of <478 K, 460 K, and 476 K from the Bi, Tl, and In cosmothermometers, respectively.[49] Leo Alaerts and Edward Anders (University of Chicago) re-examined previously published data on the release of the highly volatile elements Bi, Tl, and In from heated chondrites.[50] They were interested in the roles played by three different release processes: desorption (the opposite of adsorption; the loss of weakly held surface atoms as the result of reduced pressure or increased temperature), volume diffusion (the migration or flow of atoms or molecules through the volume or bulk of a material), and decomposition of the host mineral. Alaerts and Anders reported that the main release of Tl and Bi in Abee occurred by desorption between 400 and 700 °C, accompanied by slower acting volume diffusion. They concluded that the main release of In above 600 °C was probably also due to volume diffusion.

Table 15.1. Major and minor elements in the Abee meteorite.

Method	Si (mg)	Al (mg)	Fe (mg)	Ca (mg)	Na (mg)	Mg (mg)	Mn (mg)	Ni (mg)	Cr (mg)
INAA and RNAA		5.7–9.7	185–575	5.2–37.4	4.12–9.61	56–145	0.58–5.30	6.7–41.2	1.66–47
X-ray spectrometry									
INAA		8.4	369		8.6		3.81		3.2
Spectrophotometry, etc.			324				1.28	2.03	4.96
Mass spectrometry, etc.			291	6.74	8.78	103.6			
INAA and RNAA			305					17.8	
RNAA								23.3	
NAA	168								
Spectrometry	184.6	11.3							
Spectrophotometry, etc.			222				2.0	16.6	
NAA			222					17	
Gas chromatography									
Mass spectrometry									
Mass spectrometry									
NAA and RNAA		7.7	313		6.7		1.8	19.1	3.5
Mass spectrometry									
Mass spectrometry									
Spectrophotometry									

Note: Concentrations are given in units per gram of meteorite. Method abbreviations are as follows: NAA, neutron activation analysis; INAA, instrumental neutron activation analysis; RNAA, radiochemical neutron activation analysis.

R.A. Schmitt, R.H. Smith, J.E. Lasch, and their co-workers (General Atomic, San Diego) made a comprehensive investigation of the abundances and isotopic ratios of rare earth elements (REE) plus scandium (Sc) and yttrium (Y) in Abee and numerous other meteoritic and terrestrial specimens by neuton activation analysis (NAA).[51] Fifteen years later, N.M. Evensen, P.J. Hamilton, and R.K. O'Nions (Lamont-Doherty Geological Observatory, Palisades, New York) analyzed 15 chondrites for REE abundances by isotope dilution mass spectrometry.[52] They analyzed three separate samples of Abee, and the results were remarkable for their variation: differences in elemental abundances within the Abee meteorite were similar to those between any two of the 14 others, which encompassed carbonaceous, ordinary, and enstatite chondrites. The isotopic and elemental abundance of ytterbium (Yb) in Abee and several other meteorites was measured by Malcolm McCulloch, Kevin Rosman, and John De Laeter at the Western Australian Institute of Technology (Table 16.2b).[53] Using solid

Table 15.1. *(continued)*

Method	S (mg)	C (mg)	P (mg)	K (µg)	Ti (µg)	N (µg)	Reference
INAA and RNAA				370–1080			Sears et al. 1983
X-ray spectrometry							von Michaelis et al. 1969
INAA							Schmit et al. 1972
Spectrophotometry, etc.			2.13				Greenland and Lovering 1965
Mass spectrometry, etc.			2.35	1040			Shima and Honda 1967b
INAA and RNAA							Rambaldi and Cendales 1980
RNAA							Hertogen et al. 1983
NAA							Vogt and Ehmann 1965
Spectrometry				870			Rowe et al. 1963
Spectrophotometry, etc.	61.2						Dawson et al. 1960
NAA							Imamura and Honda 1976
Gas chromatography		3.7					Belsky and Kaplan 1970
Mass spectrometry		4.2					Grady et al. 1986
Mass spectrometry	66.7						Kaplan and Hulston 1966
NAA and RNAA							Baedecker and Wasson 1975
Mass spectrometry						254.2–853.4	Thiemans and Clayton 1983
Mass spectrometry						531	Grady et al. 1986
Spectrophotometry					494		Greenland and Goles 1965

source mass spectrometry, McCulloch et al. found a mean Yb concentration of 0.16 ppm in Abee; this is both higher and lower than values reported elsewhere, a fact they attributed to the inhomogeneity of Abee. Noboru Nakamura and Akimasa Masuda (Science University of Tokyo) determined REE abundances in Abee and several other chondrites accurately by isotope dilution mass spectrometry.[54] Two samples of Abee yielded different amounts of Yb (0.575 vs. 0.184 ppm), further testimony to the heterogeneity of this meteorite.

In 1980, R.M. Frazier and W.V. Boynton (University of Arizona) presented a paper at the 43rd Annual Meeting of the Meteoritical Society in La Jolla, California, on their determination of Yb and other trace and major elements in four samples from three clasts in Abee by both INAA and radiochemical neutron activation analysis (RNAA).[55] However, none of their samples had the large Yb anomaly observed by Nakamura and Masuda. Five years later, Frazier and Boynton presented full details of their multielemental analyses of Abee clasts,

Table 15.2a. Trace elements in the Abee meteorite.

Method	Co (µg)	Mo (µg)	Sr (µg)	Ba (µg)	Zr (µg)	Sc (µg)	V (µg)	Cu (µg)	Zn (µg)	Se (µg)	Ga (µg)	Ge (µg)
INAA and RNAA	243–2070					1.8–11.7	24–208		124–660	16–61	5.5–43	39–142
INAA	870					7.5	190					
NAA	892								454	34	17.4	
NAA	887								106	23.3	20.3	
NAA	878								419	28.1	20.7	
NAA									519	24.5		
Spectrophotometry, etc.	878		32.4	34.6		7.5	163	172	112			46
Mass spectrometry, etc.			8.3	1.6								
NAA	880							202	320	34	14	
INAA and RNAA	892							300			16.5	54.6
RNAA									559	25.6		39.4
NAA					35							
NAA								224–435	210–348			
Spectrophotometry								730				
Spectrophotometry												
Spectrophotometry												29.3
Spectrophotometry	700			500		<100	100					
NAA				1.8								
NAA	900	4.37*										
Gamma-ray spectrometry												
NAA												
NAA												
NAA and RNAA	920					4.4	58				17.7	48
NAA											71*	172*

Note: Concentrations are given in units per gram of meteorite. Method abbreviations are as follows: NAA, neutron activation analysis; INAA, instrumental neutron activation analysis; RNAA, radiochemical neutron activation analysis.

*In metal phase.

matrix, magnetic and non-magnetic separates, and dark inclusion samples.[56] The REE were analyzed using RNAA, and INAA was used for the other elements.

In 1970, Arthur J. Loveless, then a graduate student at the University of British Columbia in Vancouver, measured the isotopic composition of gadolinium (Gd), europium (Eu), and samarium (Sm) in the Abee meteorite using mass

Table 15.2a. *(continued)*

Method	As (µg)	F (µg)	Cl (µg)	Br (µg)	Rb (µg)	Ru (µg)	Sn (ng)	Rh (ng)	Hf (ng)	Reference
INAA and RNAA	1.4–8.8			2.5–5.5						
INAA										Sears et al. 1983
NAA										Schmitt et al. 1972
NAA										Ikramuddin et al. 1976
NAA										Ikramuddin et al. 1979
NAA										Biswas et al. 1980
Spectrophotometry, etc.					2.4					Laul et al. 1973
Mass spectrometry, etc.		280								Greenland and Lovering 1965
NAA					4.85					Shima and Honda 1967b
INAA and RNAA	2.4									Binz et al. 1974
RNAA	3.5					0.96				Rambaldi and Cendales 1980
NAA				3.63	2.61					Hertogen et al. 1983
NAA									170	Setser and Ehmann 1964
Spectrophotometry										Greenland and Goles 1965
Spectrophotometry										Nishimura and Sandell 1964
Spectrophotometry						1.02	250			Sen Gupta 1968
Spectrophotometry							880			Shima 1964
NAA										Dawson et al. 1960
NAA										Reed et al. 1960
Gamma-ray spectrometry										Imamura and Honda 1976
NAA		228								Reed 1964
NAA			432	3.5						Reed and Allen 1966
NAA and RNAA			500							Von Gunten et al. 1965
NAA										Baedecker and Wasson 1975
										Fouche and Smales 1966

spectrometry.[57] He observed no significant Gd, Eu, or Sm isotopic anomalies. Some isotopes of these elements have very large thermal neutron cross sections, which means they have a very high probability of absorbing thermal (slow) neutrons, thereby becoming heavier and possibly unstable isotopes. A.J. Loveless, S. Yanagita, H. Mabuchi, M. Ozima, and R.D. Russell subsequently confirmed

Table 15.2b. Trace elements in the Abee meteorite.

Method	Cd (ng)	Sb (ng)	Cs (ng)	La (ng)	Sm (ng)	Eu (ng)	Yb (ng)	Lu (ng)	Nd (ng)	Gd (ng)	Ce (ng)	Er (ng)	Pd (ng)
INAA and RNAA	210–1560	220–420		140–1240	75–990								
NAA			265										
NAA			156										
NAA	825		260										
NAA	774		216										
NAA				150	95	45	94[a]	19[b]	240	160[c]	480[d]	131[e]	
Mass spectrometry							160						
Mass spectrometry							184–575						
Mass spectrometry, etc.			240	330	180	58	200	23	570	310	710	180	
NAA	440	231											
INAA and RNAA		197											940
RNAA	950	215											1000
NAA	3300												
NAA													
Spectrophotometry													450
Fisson track													
NAA													
Mass spectrometry													
NAA and RNAA													
NAA													
INAA and RNAA				252.5	142.3	51.27	167.9	25.50	480	207	626		

Note: Concentrations are given in units per gram of meteorite. Method abbreviations are as follows: NAA, neutron activation analysis; INAA, instrumental neutron activation analysis; RNAA, radiochemical neutron activation analysis.

[a] ^{174}Yb.

[b] ^{176}Lu.

[c] ^{158}Gd.

[d] ^{142}Ce.

[e] ^{170}Er.

[f] Mean of three samples.

the absence of Gd, Eu, and Sm anomalies and proposed reasons why the Gd, Eu, and Sm atoms in Abee show no evidence of the irradiation to which Abee was exposed in space.[58] They suggested that Abee may have been shielded within a parent body with a diameter much greater than 50 m; there may have been

Table 15.2b. (continued)

Method	Os (ng)	Ir (ng)	Pt (ng)	Au (ng)	Ag (ng)	Bi (ng)	Tl (ng)	In (ng)	Reference
INAA and RNAA	560–690	130–660		103–840					Sears et al. 1983
NAA					386	134	98.1	110	Ikramuddin et al. 1976
NAA					295	15.9	7.22	30	Ikramuddin et al. 1979
NAA					289	112	93.9	96.4	Biswas et al. 1980
NAA					316	139	105	86	Laul et al. 1973
NAA									Schmitt et al. 1963
Mass spectrometry									McCulloch et al. 1977
Mass spectrometry									Nakamura and Masuda 1973
Mass spectrometry, etc.									Shima and Honda 1967b
NAA				430–930		87.8	77	80.2	Binz et al. 1974
INAA and RNAA	606	550	1230	340					Rambaldi and Cendales 1980
RNAA	771	677		408	285	71	79	131	Hertogen et al. 1983
NAA									Schmitt et al. 1963
NAA		83	1300	370					Baedecker and Ehmann 1965
Spectrophotometry	590	110	730						Sen Gupta 1968
Fisson track						44[f]			Woolum et al. 1979
NAA						80	96		Reed et al. 1960
Mass spectrometry	870	510	1190	430		99	66		Hintenberger et al. 1973
NAA and RNAA		0.50		0.44				57	Baedecker and Wasson 1975
NAA								56	Fouche and Smales 1966
INAA and RNAA									Frazier and Boynton 1985

insufficient hydrogen in the solar nebula to slow down the neutrons so that they could be absorbed, or Abee did not experience a sufficiently intense irradiation.

L. Greenland and J.F. Lovering (Australian National University) determined the abundances of 16 trace elements in a total of 50 meteorites, including Abee, by various analytical techniques.[59] They reported their results with the usual caveat, "[on account of their inhomogeneity] the sampling problem of chondrites is insurmountable and it should be recognized that any analysis is necessarily non-representative."[60] Masako Shima and Masatake Honda (University of Tokyo) studied the distribution of the alkali trace elements lithium (Li), sodium (Na), potassium (K), rubidium (Rb), and Cs; the alkaline earth elements magnesium (Mg), Ca, strontium (Sr), and barium (Ba); REE; and Fe and phosphorus (P) in Abee and two other Alberta meteorites.[61] Claude W. Sill and Conrad P. Willis (U.S. Atomic Energy Commission, Idaho Falls, Idaho) determined the concentration of the element beryllium (Be) in Abee (0.02 ppm) using a fluorometric technique.[62] At Purdue University, the team of C.M. Binz, R.K. Kurimoto, and M.E. Lipshutz used neutron activation analysis to determine gold (Au), Co, arsenic (As), copper (Cu), antimony (Sb), Ga, Se, Te, Zn, cadmium (Cd), Bi, Tl, and In levels in eight enstatite chondrites and three samples of Abee.[63] They found that, in enstatite meteorites, the abundances of the first seven elements vary by factors of 2–7, that of Te by a factor of 12, and those of the last five elements vary by much larger factors of 30–150. This abundance pattern is unlike those in other primitive chondrites, the carbonaceous and unequilibrated ordinary chondrites. Binz and his co-workers also found patterns of inter-element relationships (e.g., as Tl increases, so does Bi) that indicate the conditions in which enstatite chondrites formed were substantially different from those of the carbonaceous and unequilibrated ordinary chondrites.

Denis M. Shaw (McMaster University) stated that, in studies of element distribution in meteorites, "the conclusions tend to be limited to the pairs of elements studied and more complex inter-relationships are less clearly delineated."[64] Shaw perfomed a factor analysis on what he called the "Purdue University data" (e.g., Binz et al., see note 63; Ikramuddin et al., see notes 45 and 46) and the "University of Chicago data" (e.g., Laul et al., see note 48; Alaerts and Anders, see note 50), looking for relationships among groups of elements. He found that, in the first "Purdue" set of data, the strongest association was between Ga, Se, Te, Zn, Cd, Bi, Tl, and In: the abundance of any one of these elements affects the abundance of the other seven. By contrast, the behaviour of Sb is different, and that of As differs from both; Co shows behaviour shared between the first two groups. The "Chicago" data included a different set of elements, but factor analysis yielded similar results. Shaw's factor analysis showed "that the E4 [chondrites (e.g., Abee)] are clearly distinguished from the others [E5 and E6]." In 1966, K.F. Fouche and A.A. Smales (Atomic Energy Research Establishment, Harwell, Great Britain) determined the distribution of Ga, germanium (Ge) and In in the magnetic and non-magnetic minerals of Abee and 20 other enstatite and ordinary chondrites by NAA.[65] They said that their results supported the earlier suggestion by E. Anders that enstatite chondrites should be subdivided into two groups, namely Type I and Type II.[66] Ermanno R. Rambaldi and Miguel Cendales (Max-Planck-Institut für Chemie, Mainz, Germany) measured the concentration of some 15 siderophile ("iron-loving") elements in the magnetic and non-magnetic minerals of Abee and an E6 chondrite, Hvittis.[67] Their results, obtained by INAA, indicated that with the exception of copper, tungsten and iron, all the elements (Ni, Co, Pd, Au, As, Sb, Ga, Ge, Ir, Os (osmium), Re, Pt, and Ru) are strongly concentrated in the metal, or magnetic, phase of both meteorites. However, Abee lost more Ir (and Os, Pt, etc.) during agglomeration and accretion than Hvittis. Rambaldi and Cendales concluded that E4–E5 and E6 chondrites formed under different conditions.

Hardly a scientific paper fails to comment on the uniqueness of enstatite chondrites. A report by the University of Chicago research team of Jan Hertogen, Marie-Josée Janssens, H. Takahashi, John W. Morgan, and Edward Anders stated that "enstatite (or E) chondrites are truly a breed apart" and goes on to review their origin.[68] Hertogen et al. discussed four different factors that may have a bearing on the origin of E chondrites: pressure, formation location, reducing conditions, and metamorphism vs. condensation. Several other researchers have suggested that E chondrites condensed at higher pressures (~1 atm; 100 kPa) than other chondrites (10^{-4}–10^{-6} atm), but Hertogen et al. pointed to the low abundance of primordial noble gases in E chondrites as evidence against this argument. They also dismissed the case for E chondrites having formed in the innermost solar system (e.g., see Kempton), perhaps even less than 1 AU from the Sun.[69] The E chondrites apparently formed in a gas depleted of oxygen, one explanation for which is the freezeout of water, but as Hertogen et al. pointed out: "water ice condensed in the solar nebula only at and beyond the orbit of Jupiter, whereas E asteroids occur well inward, at 1.9–2.7 AU."[70] They judge that current data are still inadequate to decide whether the depletion of volatile elements in some E chondrites is due to incomplete nebular condensation or to volatilization by thermal metamorphism.

J.L. Setser and W.D. Ehmann (University of Kentucky, Lexington) measured levels of the trace elements zirconium (Zr) and hafnium (Hf) in Abee by NAA, reporting a Zr/Hf ratio (by weight) of 206 in Abee, much higher than the average in terrestrial rocks.[71] L. Greenland and G.G. Goles (University of California, San Diego) measured Cu and Zn abundances in two samples of Abee by RNAA.[72] Zinc was also determined, this time spectrophotometrically, in Abee by M. Nishimura and E.B. Sandell (University of Minnesota, Minneapolis).[73] They found that Zn was concentrated in the troilite (sulphide) phase (3210 ppm), with lower levels in the silicate (190 ppm) and metal (280 ppm) phases. Cadmium (Cd) is an interesting element, having eight stable isotopes with atomic masses ranging from 106 to 116, and is of considerable interest to meteoriticists and cosmochemists. R.A. Schmitt, R.H. Scott, and D.A. Olehy (General Atomic Corporation, San Diego, California) measured the abundances of Cd in over 30 meteorites by NAA.[74] They found the highest level in Abee (3.3 ± 0.1 ppm, equivalent to an atomic abundance of 4.63 Cd atoms/10^6 Si atoms). This was ~3 times higher, on a per unit weight basis, than Cd in carbonaceous chondrites and ~44 times higher compared with Cd in ordinary chondrites.

In 1965, P.A. Baedecker and W.D. Ehmann made NAA determinations of Au, Ir, and Pt in a wide range of meteorites, tektites, and a number of terrestrial materials ranging from deep-sea sediments to granite rocks.[75] Ten years later, Baedecker and J.T. Wasson made more precise measurements of Au, Ir, In, Ga, and Ge by RNAA and 10 additional elements by NAA.[76] The concentrations of the six platinum group metals—ruthenium (Ru), rhodium (Rh), palladium, osmium, iridium, and platinum) were measured spectrophotometrically in several meteorites by J.G. Sen Gupta of the Geological Survey of Canada.[77]

Masako Shima, then at the University of California, San Diego, determined the distribution of germanium and tin (Sn) in some 30 meteorites colorimetrically, using a Beckman DU spectrophotometer (see Chapter 8).[78] James R. Vogt and William D. Ehmann determined the silicon abundances in over 100 meteorites by fast neutron activation analysis. They found Abee to contain 16.8% Si by weight;[79] this was somewhat less than the value determined in the initial analysis by Dawson et al. (q.v.).

Boron (B) is a moderately volatile element whose abundance in meteorites can shed information on the nature of the primordial solar nebula and its condensation history. David Curtis, Ernest Gladney, and Edward Jurney (Los Alamos Scientific Laboratory) determined the B concentration in Abee and some 50 other meteorites by PGNAA (Table 15.2c).[80] Mingzhe Zhai and Denis M. Shaw (McMaster University) used the same method to determine the average B concentration of the Abee meteorite (1.19 ppm, the highest level among the three enstatite chondrites tested).[81] Zhai and Shaw, in collaboration with Eizo Nakamura and Toshio Nakano (Okayama University), later made two measurements of the isotopic ratio $^{11}B/^{10}B$ in Abee.[82]

M.T. Murrell and D.S. Burnett at the California Institute of Technology measured the distribution and concentrations of thorium (Th) and uranium (U) in Abee and several other enstatite meteorites by fission track radiography (see J.N. Goswami, below).[83] They found that oldhamite and niningerite are the main U-bearing minerals in Abee, whereas oldhamite is the principal Th bearer. Hideki Masuda, Masako Shima, and Masatake Honda (University of Tokyo) measured the distribution of U and Th by chemical dissolution and fractionation of the non-magnetic phases in powdered samples of four chondrites including Abee.[84] The dried fractions were irradiated and U and Th determined by NAA. Dorothy S. Woolum, Linda Bies-Horn, D.S. Burnett and L.S. August also used fission track radiography to map the distribution of Bi and lead (Pb) in polished sections of Abee and four other enstatite chondrites.[85] Woolum and her coworkers found that Bi and Pb are not concentrated in metal or sulphide grains and are comparatively homogeneously distributed. This is consistent with Abee and the other meteorites having formed by condensation from the solar nebula.

C.L. Smith, J.R. De Laeter, and K.J.R. Rosman (Western Australian Institute of Technology) determined the abundance of tellurium (Te) in 25 chondrites, 3 achondrites, a tektite and 12 standard terrestrial rocks with a stable isotope dilution technique using solid source mass spectrometry.[86] They believed their data represented a significant improvement in accuracy in Te measurements. However, conflicting interpretations of data from other determinations of Te isotopic composition in Abee later prompted Smith and De Laeter to make new mass spectrometric analyses of samples of Abee on two independent and different instruments, using rigourous measures to avoid sample contamination, careful instrument calibration procedures, and the isotope dilution technique.[87] Smith and De Laeter compared their new data with a standard terrestrial Te sample and reported no isotopic anomalies can be distinguished within the error limits.

Mass spectrometric analyses of calcium present special difficulties because of the large mass difference of the isotopes, ranging from ^{40}Ca to ^{48}Ca, and the low abundance of all Ca isotopes other than ^{40}Ca. To counteract these problems W.A. Russell, D.A. Papanastassiou, and T.A. Tombrello (California Institute ot Technology) used a so-called "double-spike" technique in their analysis of numerous terrestrial, lunar, and meteoritic samples.[88] Russell et al. tested two different samples of Abee to check the possibility that it might contain fractionated (isotope enriched or depleted) Ca in oldhamite (CaS). They found no instances of Ca fractionation greater than 2.5‰; furthermore, meteorites (including Abee), lunar samples, and terrestrial samples all show this same small range of fractionation, indicating remarkable initial isotopic homogeneity in different solar system objects. F.R. Niederer (Universität Bern), D.A. Papanastassiou, and G.J. Wasserburg (California Institute of Technology) used mass spectrometry to determine the absolute abundances of Ti isotopes in Abee and several other meteorites.[89]

Table 15.2c. Trace elements in the Abee meteorite.

Method	Pb (ng)	Tb (ng)	Dy (ng)	Hg (ng)	Pr (ng)	Ho (ng)	Tm (ng)	Re (ng)	Y (ng)	Th (ng)
Mass spectrometry, etc.			310							
NAA										
NAA										
NAA										
NAA										
NAA	25		160		54	40	14		1020	
Fluorometry										
NAA										
INAA and RNAA								44.5		
RNAA								64.0		
NAA										
PGNAA										
Fission track										910
Mass spectrometry										
Mass spectrometry										
NAA										
NAA				4.0						
Mass spectrometry	1980		420					46		23
NAA										
INAA and RNAA		36.45	259.0		80	53.2	24.9			

Note: Concentrations are given in units per gram of meteorite. Method abbreviations are as follows: NAA, neutron activation analysis; INAA, instrumental neutron activation analysis; RNAA, radiochemical neutron activation analysis.

G.G. Goles and E. Anders (University of California, La Jolla) determined the iodine (I), Te, and U content of some dozen chondrites, including Abee, by NAA.[90] Their results ranged so widely—levels of I varying by factors of up to 100 and Te and U up to 15 times—that they were cautious in deriving mean abundances or "cosmic" abundances from their data, echoing the concerns of Greenland and Lovering and others.

The concentration of noble gases—helium (He), neon (Ne), argon, (Ar), krypton (Kr), xenon (Xe), and radon (Ra)—in meteorites is so low that radioactive decay or nuclear reactions induced by cosmic rays produce measurable isotopic variations.[91] The study of noble gases in meteorites yields information on the early history and evolution of the solar system. Craig Merrihue

Table 15.2c. *(continued)*

Method	U (ng)	Te (ng)	W (ng)	I (ng)	Be (ng)	B (ng)	Reference
Mass spectrometry, etc.							Shima and Honda 1967b
NAA		2540					Ikramuddin et al. 1976
NAA		1210					Ikramuddin et al. 1979
NAA		2460					Biswas et al. 1980
NAA							Laul et al. 1973
NAA							Schmitt et al. 1963
Fluorometry					20		Sill and Willis 1962
NAA		2700					Binz et al. 19774
INAA and RNAA			138				Rambaldi and Cendales 1980
RNAA	9.1	2410					Hertogen et al. 1983
NAA						1000	Curtis et al. 1980
PGNAA						1190	Zhai and Shaw 1994
Fission track	260						Murrell and Burnett 1982
Mass spectrometry							Smith et al. 1977
Mass spectrometry							Smith and De Laeter 1986
NAA	15	2200		145			Goles and Anders 1962
NAA							Reed et al. 1960
Mass spectrometry	6.8		370				Hintenberger et al. 1973
NAA		2750		≤180			Reed and Allen 1966
INAA and RNAA							Frazier and Boynton 1985

(University of California, Berkeley), in an appendix to his 1966 paper on the Bruderheim meteorite, briefly reported on the abundance of Kr and Xe in Abee.[92] W.B. Clarke and H.G. Thode (McMaster University) made measurements of the isotopic composition of krypton in Abee by heating meteorite samples in a vacuum and analyzing the gases driven off in a mass spectrometer.[93] Abee and some of the other meteorites tested have a Kr isotopic composition more like that of the terrestrial atmosphere than Bruderheim. P.M. Jeffrey and J.H. Reynolds (University of California, Berkeley) investigated the origin of the "excess" levels (compared with the cosmic abundance) of the xenon isotope ^{129}Xe by heating a 1.5 g sample of Abee, in 100° steps, between 300 °C and 1500 °C.[94] The gases released during each step were analyzed in a

Table 15.2d. Noble gas isotope concentration in the Abee meteorite.

Method	^{3}He (cm^3 STP/g, ×10^{-8})[a]	^{4}He (cm^3 STP/g, ×10^{-8})	^{22}Ne (cm^3 STP/g, ×10^{-8})	^{36}Ar (cm^3 STP/g, ×10^{-8})
Mass spectrometry	8.2–12.2	549–1440	0.38–2.87	7.8–46.8
Mass spectrometry			1.0–4.7	30–1020

[a]STP, standard temperature and pressure (273.15 K and 100 kPa, respectively).

[b]Measurements were made in the oldhamite phase.

mass spectrometer. Jeffery and Reynolds concluded that the excess ^{129}Xe originated from *in situ* decay of the radioactive isotope ^{129}I. B. Srinivasan, R. Lewis, and E. Anders (University of Chicago) likewise examined samples of Abee and Allende (a carbonaceous chondrite) by stepwise heating, in an attempt to relate the noble gases released at different temperatures to specific minerals.[95] Srinivasan et al. tentatively identified the mineral *djerfisherite* as the host for the noble gases released when Abee was heated to 800 °C; troilite (at 1100 °C); and silicates (~1300 °C). Jane Crabb and Edward Anders, also at the University of Chicago, took the next logical step in investigating the siting of noble gases in Abee and five other enstatite chondrites by analyzing fractions separated by density, grain size, and chemical resistance.[96] Abee proved to be particularly difficult to disaggregate and Crabb and Anders found that the highest concentrations of Ne, Kr, and Xe in Abee were in the residue left after treatment with hydrofluoric and hydrochloric acids (Table 15.2d). Most of the Ar in Abee is found in the residue left after dissolution in HCl, followed by sedimentation. At least some of this chemically resistant Abee residue must consist of diamonds.

The concentrations in Abee of the trace heavy elements barium (Ba), mercury (Hg), thallium, lead, bismuth, and uranium were determined by G.W. Reed, K. Kigoshi, and A. Turkevich (University of Chicago) using RNAA.[97] H. Hintenberger, K.P. Jochum and M. Seufert (Max-Planck-Institut für Chemie, Mainz, Germany) report that "it is possible to recognize at least the main class of a meteorite from its content of the heavy elements."[98] Using spark source mass spectrometry, they determined concentrations of trace heavy elements in four Antarctic meteorites, Yamato (a), (b), (c), and (d) and seven other meteorites including Abee. A close match between these elements in Abee and in Yamato (a) confirmed the preliminary classification of the latter as an enstatite chondrite. Keiko Imamura and Masatake Honda of the Institute for Solid State Physics at the University of Tokyo measured the concentration of molybdenum (Mo) and Co in the separated metal, silicate, and troilite phases of Abee using

Table 15.2d. *(continued)*

Method	^{84}Kr (cm³ STP/g, ×10⁻¹²)	^{132}Xe (cm³ STP/g, ×10⁻¹²)	Reference
Mass spectrometry	205–2160	86–970	Wacker and Marti 1983
Mass spectrometry	1700–58,000[b]	740–50,000[b]	Crabb and Anders 1982

NAA.[99] They report Mo is concentrated in the metal phase (4.31 ppm) with lesser levels in silicates (0.65 ppm) and troilite (2.7 ppm); the distribution of Co is more disproportionate still: metal phase (3700 ppm), silicates (26 ppm), and troilite (235 ppm).

The halogens—fluorine (F), chlorine (Cl), bromine (Br), iodine (I), and astatine (At)—are also of considerable interest to meteoriticists. George W. Reed (Argonne National Laboratory, Argonne, Illinois) used a neutron activation technique to measure the F content in Abee and several stony meteorites.[100] Reed and Ralph O. Allen (University of Chicago) measured the Cl, Br, and I contents in samples of several ordinary, carbonaceous, and enstatite chondrites by NAA.[101] They found that Cl is heterogeneously dispersed in all the meteorites tested, except Abee. In contrast to Cl, the Br and I contents of all chondrites tested, including Abee, did not vary greatly. Reed and Allen also determined the Te and U content of Abee. H.R. von Gunten, A. Wyttenbach, and W. Scherle (Eidg. Institut für Reaktorforschung, Würenlingen, Switzerland) also measured the Cl content in Abee and a few other stony meteorites by NAA.[102] Their mean result for two Abee samples, 500 ± 6 ppm, was the highest among the 11 meteorites tested.

Stable Isotopes

Abee Consortium members Mark H. Thiemens and Robert N. Clayton, at the Enrico Fermi Institute (University of Chicago), determined the nitrogen (N) content and isotopic ratios ($^{15}N/^{14}N$) of three clasts from Abee.[103] Nitrogen was extracted from each sample of crushed clast as it was heated stepwise to 1500 °C under vacuum conditions; the isotopic composition of the N extracted at each temperature step was measured with a mass spectrometer. Thiemens and Clayton found that the greatest amount of N was released at a temperature of 1300 °C, suggesting that it had been incorporated into the meteorite at high temperature in the solar nebula. They also found that "the isotopic ratio in Abee is one of the most ^{15}N-depleted observed in all meteorites."[104]

Monica M. Grady, I.P. Wright, L.P. Carr, and C.T. Pillinger (Open University, Milton Keynes, England) measured carbon (C) and N abundance and isotopic isotopic composition ($^{13}C/^{12}C$ ratio, or $\delta^{13}C$, = −13.4‰) in Abee.[105,106] They also reported a total N content (531 ppm) and a ^{15}N enrichment in three separate stepped pyrolyses (heating in the absence of oxygen) of Abee, with $\delta^{15}N$ values of 269‰ (400–500 °C), 230‰ (400–600 °C), and 350‰ (500–550 °C). They found this evidence of isotopically heavy N only in Abee and not in the other enstatite chondrites.

In 1970, T. Belsky and I.R. Kaplan (University of California, Los Angeles) detected and identified several hydrocarbons trapped between crystal boundaries in several meteorites including Abee.[107] They heated crushed bulk samples of the meteorites and analyzed the gases released using gas chromatography. Several compounds ranging from methane (147 ppb) to benzene (50 ppb) were found in Abee. Belsky and Kaplan also measured the total carbon (0.37%) in Abee and its isotopic composition ($\delta^{13}C$ = −8.1‰). Jongmann Yang and Samuel Epstein (California Institute of Technology, Pasadena) later reported measuring the isotopic composition of carbon and hydrogen in some 20 meteorites by mass spectometry.[108] They found meteoritic $\delta^{13}C$ (Abee = −12.4‰) to be well within the range observed on Earth. On the other hand they found the magnetic and non-magnetic phases of Abee were both quite depleted in deuterium (δ^2H; Abee = −340‰) relative to terrestrial values.

The year 1953 was an epochal year for molecular biology. In the April 2 issue of *Nature*, James Watson and Francis Crick published their model for the structure of deoxyribonucleic acid (DNA) and, more relevant to our story, in the 15 May issue of *Science*, Stanley L. Miller reported his classic experiment which would change the approach of scientific investigation into the origin of life.[109] Miller was a University of Chicago graduate student working in Harold C. Urey's chemistry laboratory. Miller and Urey designed an experiment to replicate conditions believed to have existed on the early, prebiotic Earth. Their apparatus consisted of a closed system of glass flasks containing a mixture of methane, ammonia, hydrogen, and water which was repeatedly zapped by electrical discharges (representing lightning). The experiment yielded organic compounds including amino acids, the building blocks of life, showing how organic matter might have been synthesized in the early solar system.[110]

The discovery that Abee contained diamonds was, in some ways, not surprising, for they were first reported in a meteorite over a century ago. However, the diamonds found in Abee by S.S. Russell, C.T. Pillinger, J.W. Arden, M.R. Lee, and U. Ott held different surprises.[111] The diamonds found in virtually all types of primitive carbonaceous and ordinary chondrites, as well as some enstatite chondrites have C and N isotopic ratios characterisitic of interstellar space, indicating that they formed out of presolar material; however, Abee diamonds have typical solar system isotopic compositions for C, N, and Xe and, presumably, originated in the solar nebula. In a subsequent paper presented at the 1992 Annual Meeting of the Meteoritical Society in Copenhagen, U. Ott, H.P. Löhr and J.W. Arden comment "the presence of cosmogenic ^{4}He and ^{21}Ne [is] unambiguous proof of the indigenous [solar system] nature of the Abee diamonds."[112] How did these diamonds form? It is unlikely that they formed the same way as terrestrial diamonds because their parent bodies, with diameters of no more than a few hundred kilometres, would not reach the required high pressures such as are found deep in the Earth. It is believed that the diamonds in the Canyon Diablo iron meteorite were formed by the shock of impact, but this mechanism is ruled out in Abee, which had a comparatively gentle landing. Russell et al. suggest that the Abee diamonds, which are ~100 nm in size and represent about 100 ppm of Abee by mass, were formed by low-pressure synthesis, possibly by chemical vapour deposition (CVD) or as a result of radiation by high-energy particles.[113]

According to Peter Deines (Pennsylvania State University, University Park) and Frans Wickman (Stockholms Universitet, Sweden) enstatite chondrites may contain carbon in several forms, for example, graphite, cohenite, hydrocarbons, carbides, and carbonates.[114] Deines and Wickman measured the isotopic composition of the total carbon in Abee and several other enstatite chondrites plus an enstatite achondrite. They found that both carbon content and isotopic composition—there are two naturally occurring stable isotopes, ^{12}C (~99%) and ^{13}C (~1%)—varied widely between the meteorites tested and within each meteorite; one sample of Abee "contained so little carbon that we could not use it for our measurements." However, in general, the ^{13}C content of enstatite chondrites increases with petrographic type: E4 (e.g., Abee) < E5 < E6. Deines and Wickman suggest that if differences occurred in the C/O ratio within the solar nebula during the condensation of material forming enstatite chondrites, "then under more oxidizing conditions carbon would precipitate from CO as graphite and show a relative ^{13}C enrichment, while under more reducing conditions [as experienced by Abee] carbon would be precipitated in the form of cohenite showing a relative depletion in ^{13}C with respect to the gas reservoir."[115]

Hugh Taylor, Michael Duke, Leon Silver, and Samuel Epstein at the California Institute of Technology in Pasadena made an oxygen (O) isotope study of minerals in over 30 meteorites using mass spectrometry. They found that for stony meteorites the ^{18}O/^{16}O ratio for coexisting minerals increased in the following sequence: olivine < pyroxene < plagioclase < free silica (e.g., quartz).[116] Taylor and his colleagues measured only the pyroxene [18]O enrichment in Abee ($\delta^{18}O = 6.0 \pm 0.1$‰). Concurrently, J.H. Reuter, S. Epstein, and H. Taylor reported that the variation in the ^{18}O/^{16}O ratios of some 30 chondrites, including Abee, is very small compared with the variation that is observed for terrestrial rocks.[117]

Sulphur (S) is a common and often important constituent of many meteorites. M. Shima and H.G. Thode (McMaster University, Hamilton) made two separate determinations of sulphur isotope ratios, using mass spectrometry in each of two phases of Abee (matrix and fragments).[118] To determine the sulphur isotopic abundance and content of 20 iron and stony meteorites in more detail, I.R. Kaplan and J.R. Hulston (also from McMaster University) first chemically extracted S from the mineral phases of each meteorite, then converted this to either SO_2 or SF_6 gas for analysis by mass spectrometry.[119] Kaplan and Hulston were particularly interested in measuring the S contained by alabandite (MnS) in Abee, which had previously been identified by Dawson et al. (see note 11). They found that total S in Abee amounted to 6.67%, which was the highest level of any stony meteorite tested and exceeded only by the iron meteorites. Sulphur in Abee is concentrated in troilite (4.62%), alabandite (~1.92%), and oldhamite (0.13%).

Isotope Chronology

D.D. Bogard, D.M. Unruh, and M. Tatsumoto at NASA's Johnson Space Centre in Houston, Texas used argon-argon ([40]Ar/[39]Ar) dating and uranium-lead dating to measure the ages of three different Abee clasts as part of the consortium study (Table 15.3).[120] Samples of three Abee clasts (nos. 1,1; 2.2; 3,3 in Fig. 15.4) were irradiated with fast neutrons from a research reactor. Argon was extracted from each clast as it was heated stepwise to 1500 °C or 1550 °C; the isotopic composition of the argon extracted at each temperature step was measured with a mass spectrometer. Data from all three clasts yielded similar age determinations in the range of ~4.4–4.5 Ga, and it was not possible to resolve any age differences among the clasts. Portions of the three clast samples were also chemically analyzed for U, Th, and Pb abundances and U and Pb isotopic compositions. Unfortunately, the U/Pb and Th/Pb data did not help resolve the age and history of the Abee meteorite, probably because of sample contamination by terrestrial lead. Nevertheless, Bogard and his colleagues were able to suggest that the various Abee components experienced separate histories until brecciation (breaking up into fragments) and compaction (solidification into a meteorite) occurred, no later than 4.4 Ga ago, and that Abee experienced no appreciable subsequent heating.

Another Abee Consortium team, John F. Wacker (University of Arizona) and Kurt Marti (University of California, San Diego), measured the concentrations and isotopic compositions of the noble, or chemically inert, gases (He, Ne, Ar, Kr, and Xe) in the clasts and matrix of Abee.[121] Sample portions, after being crushed and chemically processed, were heated *in vacuo* (under vacuum conditions) to extract the noble gases. Wacker and Marti then analyzed the gases in a mass spectrometer. They were especially interested in the $^4He/^3He$ isotope ratio in helium; the $^{20}Ne/^{21}Ne$ and $^{21}Ne/^{22}Ne$ ratios in neon; and the $^{129}Xe/^{132}Xe$ ratio in xenon. Some of these isotopes are produced by the action of cosmic rays on target atoms in the meteorite when the latter is exposed to cosmic radiation in space: the future Abee meteorite was once part of a larger parent body which suffered a violent collision with another asteroid or meteorite, blasting debris into space. By measuring these so-called cosmogenic isotopes, Wacker and Marti were able to calculate the cosmic-ray exposure age of the Abee meteorite at 8 Ma.

When a uranium (^{238}U) atom in a meteorite splits, or undergoes *fission*, considerable energy is released, much of which is transferred to the two atomic fragments, called *nuclides*. These nuclides fly off in opposite directions leaving a short (~15 μm) trail of damage in the crystalline structure of the surrounding minerals. These trails, also known as *nuclear tracks*, can be etched with chemicals so that they become large enough to see under a microscope. J.N. Goswami at the Physical Research Laboratory in Ahmedabad, India, analyzed the nuclear tracks in several samples taken from different locations in the slab of meteorite being studied by the Abee Consortium.[122] Goswami observed, counted, and estimated the frequency, or density, of tracks in different locations in Abee. He found that the track density was lowest (~2×10^4 cm^{-2}) near the centre of the slab, which had received the maximum shielding from cosmic radiation and, therefore, represented the *background* level of spontaneous fission. He also found that track density increased (up to ~10^6 cm^{-2}) towards the edge of the meteorite, which had received less shielding from cosmic radiation, reflecting an *excess* contribution from induced fission. Goswami was able to deduce from this data the depth of shielding, or amount of meteoritic material that had covered Abee before it fell to Earth. He calculated that the pre-atmospheric object had a diameter of ~60 cm; therefore, Abee lost some three-quarters of its original mass from ablation during its fiery plunge. Fission tracks are annealed or "erased" at temperatures above 100 °C, so Goswami concluded that "the Abee meteorite was not reheated to high temperature during or after brecciation, which took place ~4.4 Ga ago."[123]

M.W. Rowe, M.A. Van Dilla, and E.C. Anderson (Los Alamos Scientific Laboratory) developed a nondestructive gamma-ray spectrometry method in

Table 15.3. Age estimates of the Abee meteorite.

Method	Solidification age (Ga)	Cosmic-ray exposure age (Ma)	Gas retention age (Ga)	Reference
^{3}He/^{21}Ne/^{38}Ar		8		Wacker and Marti 1983
^{40}Ar/^{39}Ar			4.4–4.5	Bogard et al. 1983
Fission track	~4.4			Goswami 1983
Rb/Sr	4.52			Shima and Honda 1967a

which the gamma-ray spectrum of a meteorite is compared with the spectra of chemically calibrated plaster replicas or "mock-ups" of the meteorite (see Chapter 8).[124] Using this method, Rowe et al. analyzed a suite of 37 chondrites and achondrites, including Abee. They found that Abee contained 0.087% K, 18.46% Si, and 1.13% Al, and they measured the level of the radioactive isotope ^{26}Al at 51 dpm/kg (disintegrations per minute per kilogram). A meteorite's *cosmic-ray exposure age*, or the length of time it has been exposed to cosmic rays in space, can be estimated indirectly from its ^{26}Al content. The production of ^{26}Al and other so-called *cosmogenic* isotopes is influenced by several factors including a meteorite's chemical composition, and its hardness or shielding from radiation and variations in cosmic ray levels. P. Englert and W. Herr (Max-Planck-Institut für Kernphysik, Heidelberg, Germany), in their study of ^{53}Mn-derived exposure ages of chondrites cite the exposure ages of several meteorites, including Abee (7.0 ± 1.9 Ma), studied by other researchers.[125] Ian Cameron and Zafer Top, at the University of New Brunswick in Saint John, measured the level of cosmogenic ^{26}Al in a piece of Abee from the Geological Survey of Canada's National Meteorite Collection (specimen no. 011710), using a gamma-ray spectrometer.[126] They measured an activity of 58.5 ± 4.5 dpm/kg, in fair agreement with Rowe et al., from which they were able to postulate an orbit for the parent body of Abee with an aphelion in the range of 2.04–2.40 AU.[127] Philip J. Cressy (NASA Goddard Space Flight Center, Greenbelt, Maryland) quotes a value of 63 ± 7 dpm/kg for the ^{26}Al level in Abee.[128]

Masako Shima and Masatake Honda (University of Tokyo) sought to determine the ages at which the parent bodies of the Bruderheim, Peace River, and Abee meteorites solidified out of the primordial solar nebula by rubidium-strontium (Rb-Sr) dating.[129] Shima and Honda comment that "the story of this enstatite chondrite [Abee] was remarkably different from the [other] two meteorites," particularly its high Rb and Cs content and its high ^{87}Sr/^{86}Sr ratio. They were able to calculate a solidification age of 4.52 Ga for Abee.

Summary

Abee may still be the most-studied Canadian meteorite. It is still in many ways a unique stone, more than 50 years after its fall. It remains, by far, the largest enstatite meteorite ever found. Scientifically, it has generated a lot of discussion. The Abee data, which include trace elements, stable isotopes, long-lived radionuclides, and shorter-lived cosmogenic nuclides, have helped set limits on the intensity of radiation around the early Sun, and in other ways constrain the addition of presolar grains to the nebula (although such things are better studied in more primitive meteorites, such as carbonaceous chondrites and some unequilibrated ordinary chondrites). In 2001, the International Mineralogical Association approved the naming of a new sulphide mineral, keilite, (after the meteoriticist Klaus Keil), found in Abee, and the type locality was designated as the site where the Abee meteorite was found.

16

Great Balls of Fire

In 1916, the Dominion Government established a geomagnetic observatory at Meanook, Alberta, about 18 km south of the town of Athabasca and 136 km north of the city of Edmonton, to measure variations in the Earth's magnetic field.[1] A basement to house Kew-type variometers (a type of magnetometer) was added to the original observatory building in 1927. Two sets of more sensitive LaCour variometers were installed in a temporary hut for the second International Polar Year (1932–1933).[2] In 1941, the LaCour variometers were moved to a basement annex added to the original building. Ten years later, a new building was built to house both the absolute instruments and the variometers, and an additional 215 hectares of land adjacent to the observatory were purchased in 1955.

The Meanook Geomagnetic Observatory continues its program of geomagnetic measurements to the present day, sharing its buildings and grounds with the University of Alberta's Meanook Biological Research Station. However, for two decades following World War II, Meanook and a sister observatory set up at Newbrook, some 40 km to the southeast, also carried out other scientific research of much greater relevance and interest to our story. The establishment of meteor observatories at these two locations was the direct result of informal discussions held in Cambridge, Massachusetts, during March 1946. Those involved included Fred L. Whipple of the Harvard College Observatory, Zdenek Kopal from MIT, and Peter M. Millman, a Canadian authority on meteor spectrography, and their conversations dealt with the value of photographic meteor research in the study of the upper atmosphere.[3] Shortly after this meeting, the Dominion Observatory in Ottawa agreed to co-ordinate Canada's participation in this research program, and Peter Millman joined the staff of the observatory to plan the development of meteor research there.

To give a satisfactory latitude spread between meteor observatories (another pair were located in New Mexico), it was necessary to locate the two Canadian meteor stations as far north as practicable. Suitable sites were found at Meanook and Newbrook, Alberta, where it was hoped that neither interference from auroral activity nor loss of night-time darkness near the period of summer

solistice would be excessive—even so, as we have seen (Chapter 15), the bright evening twilight prevented Newbrook from photographing the Abee fireball on 9 June 1952. The site chosen for the Meanook meteor station was within 75 metres of the existing geomagnetic observatory. To establish the totally new station at Newbrook, 0.8 hectares of land were purchased just south of the town (Fig. 16.1).

It was decided to equip all the new meteor observatories with the Super-Schmidt meteor camera designed and built by the Perkin-Elmer Corporation of Norwalk, Connecticut. This revolutionary camera had an aperture of 31 cm and a focal length of 20 cm, with a circular field of 55° in diameter; the complete camera and mounting weighed about 2500 kg. The camera used Kodak blue-sensitive X-ray sheet film, molded to fit tightly against a spherical plate holder and held in place during exposure by a vacuum pressure system. A two-vane shutter, rotating 3 mm above the emulsion, occulted the film 60 times per second; the effect of this is to chop any photographed meteor trail into segments which, when measured, provide information about the meteor's velocity. A special copying camera was used at the Dominion Observatory in Ottawa to copy the original spherical negative onto a flat glass plate.

The acceptance field tests of the Meanook and Newbrook cameras were carried out by Peter Millman in early 1952 in Norwalk, where the final adjustment and positioning of the heavy optical parts of the cameras was also carried out.

Fig. 16.2. Mounting the camera body at the Meanook Meteor Observatory, August 1952. Note how the roof opens; one section slides off to the left, the other to the right.

Millman continues the story (Figs. 16.2–16.4):

> To avoid the jarring which might occur in shipping by rail, the Canadian cameras were flown direct from New York to Edmonton by the Royal Canadian Air Force and then taken by special truck to Meanook and Newbrook. Small buildings, with sliding roof units and suitable reinforced concrete piers, had been constructed at these sites some time before. The two Super-Schmidts were successfully mounted during August, 1952. First test exposures at Meanook were made on the night of August 15/16 and at Newbrook on August 18/19.[4]

It was necessary to make several cold-weather modifications to the cameras before they could be operated in the winter nights of northern Alberta. Heaters were installed in the shutter motor housing, the main polar axis housing, in the dew cap, and in the main drive gear box. Even so, the cameras were seldom operated when the temperature was below -20 °C. The main purpose of the meteor photographs was to provide accurate co-ordinates for the meteor path, as measured against the background stars, and to allow the accurate measurement of the meteor's angular velocity. A secondary aim was to obtain a photometric light curve along the meteor's path and to provide information about the meteor's ablation and conditions in the upper atmosphere. Often

Fig. 16.3. The crew that mounted the Super-Schmidt cameras at the Meanook and Newbrook Meteor Observatories. From left to right: A.A. Griffin, G.A. Brealey, J.M. Grant, H.E. Cook, J. Pare, A. Page, and P.M. Millman.

photographs of the same meteor were taken at both Meanook and Newbrook, making possible the determination of heights, velocities, and decelerations by triangulation with an accuracy of about 0.1%. The photograph taken of the 5 February 1967 Vilna fireball at Meanook (see Fig. 13.2) was vital to that meteorite's recovery. In addition, the staff at Meanook and Newbrook carried out a visual watch for meteors while the cameras were operating; in one six-year period, from 1952 to 1957, 3800 meteors were visually observed at the two stations. Newbrook's most unusual achievement happened on 9 October 1957, when Art Griffin secured the first ever photograph of the first artificial satellite, Sputnik 1, taken from North America.[5]

The Meanook and Newbrook observatories were labour intensive and quite expensive to operate, requiring a permanent staff of three. Because suitable living accommodation was not available near either of the observatories, residences were built in 1951 at Meanook and 1952 at Newbrook to house the meteor personnel. In the fall of 1953, facilities at both observatories were expanded, and the office building at Meanook was enlarged in 1956–1957.

The Meanook and Newbrook observatories operated from 1952 to 1970. Many of the buildings at Meanook are still standing on beautiful park-like

Fig. 16.4. The Super-Schmidt camera mounted at the Newbrook Meteor Observatory, August 1952. The counterweights at rear are to balance the sliding roof sections.

grounds and are used by the Geomagnetic Observatory and the University of Alberta. However, the Super-Schmidt camera and other instruments were removed and shipped to Ottawa and the observatory building (Fig. 16.2) was demolished; today only the heavy concrete pier and foundations remain, abandoned and overgrown by the woods (Fig. 16.5).

Although the Newbrook Observatory was also stripped of all scientific apparatus, the Super-Schmidt camera building and the residence house were both saved from destruction and are being preserved and restored by the Newbrook Observatory Historical Society. The author was invited to visit the observatory by Hazel Humm, the society's president. The house (Fig. 16.6) was clearly once a very comfortable residence and is well maintained. Its several rooms will easily accommodate the museum, displays, and meeting facilities the historical society hopes to create. The most interesting feature of the house is a basement fallout shelter built during the height of the Cold War in response to a government regulation requiring all federal buildings to have such a facility. The separate observatory building (Fig. 16.7) is also in very good condition and could easily be restored to active use if re-equipped with a telescope.

Fig. 16.5. Ruins are all that remain today of the Meanook Meteor Observatory building that once housed the Super-Schmidt camera.

The Meanook and Newbrook observatories, with their revolutionary Super-Schmidt cameras, represented the state of the art technology in the early 1950s. A decade later, it had become apparent that the future of meteor photography lay in large-area networks of semi-autonomously operated cameras. In 1959, a network of film cameras was set up in Czechoslovakia to track Earth satellites (the forerunner of the current *European Network*), but the cameras also happened to photograph the path of a meteorite (Pribram) that was subsequently recovered. Scientists were able to calculate the orbit of the Pribram meteorite from the camera data (see Figure 10.4), the first time this had ever been done. Impressed by this result, the Smithsonian Astrophysical Observatory established and operated a network of 16 camera stations spanning the American Midwest from 1964 to 1974. This *Prairie Network* resulted in the second recovery and orbit determination of a meteorite (Lost City, Oklahoma), in 1970. Planning for a Canadian camera network began in the 1960s at the Dominion Observatory in Ottawa.[6] Site selection began in 1966, and 12 observatory locations were chosen by the spring of 1968. The first station was built at Asquith, Saskatchewan (coincidently the birthplace of Boyd Wettlaufer, discoverer of the Belly River meteorite), followed by others in Alberta (at Vegreville, Lousana, and Brooks), Saskatchewan, and Manitoba (Fig. 16.8). The Meteorite Observation and Recovery Project (MORP) was operated from a field headquarters at the University of Saskatchewan in Saskatoon. Although the first cameras were

Fig. 16.6. The residence house at the Newbrook Meteor Observatory today.

installed at Asquith late in 1968, the full network was not effectively operational on a routine basis until 1971.

Each observatory building was compact, with a floor area of only 6.5 m² (about the size of a home bathroom); heated or cooled, as necessary, by a combination heater and air conditioner and designed for nearly labour-free operation and maintenance. Each observatory was visited about twice a week by an operator who lived near the station. Local residents were trained for this in a few hours; with the aid of a manual covering routine and extraordinary procedures, these operators did an excellent job. In the entire history of the project, only one operator resigned (he was transferred by his regular employer), and one passed the job to his son.[7]

Each MORP observatory was five-sided to provide each of its five cameras with a window (Fig. 16.9). Each window was electrically heated to keep it free of ice and dew. The cameras had wide-angle lenses and used Kodak 70 mm Plus-X Pan films in 30 m rolls. Each picture exposed included, in the frame margin, a camera identifier (station letter A to L and camera number 1 to 5), three fiducial marks to allow accurate measurements of meteor trails, and a time and date display. Each observatory had, mounted on its roof, a photoelectric meteor detector that would automatically register the time of the event and advance the film. The cameras had a chopping shutter, as is usual in meteor cameras, to break up any meteor trail into segments (see Fig. 10.1). Each 30 m roll of

Fig. 16.7. The Super-Schmidt camera building at the Newbrook Meteor Observatory today, with Hazel Humm. Note the sliding roof rails; compare with Figs. 16.2 and 16.4.

film lasted 3–6 weeks and would be changed when necessary by the operator, who would then mail it to the MORP headquarters in Saskatoon. However, if as happened with the Innisfree meteorite (see Chapter 10), eyewitnesses reported a spectacular fireball, the film could be removed prematurely and rushed to Saskatoon for prompt examination. The MORP ended on 31 March 1985, after about 15 years of operation, during which just over 1000 fireballs were photographed by at least two stations. The data from these fireball photographs has been extensively studied.

Ian Halliday, Alan Blackwell, and Arthur Griffin (National Research Council of Canada) studied the data from 44 fireballs that are believed, for a variety of reasons, to have resulted in the fall of meteorites.[8] Halliday and his colleagues concluded that, by itself, the brightness of the fireball is a poor indicator of whether any meteorites survived. A low (<10 km/s) terminal velocity offers more assurance of a fall. "The survival of a meteorite is facilitated by a low [atmospheric] entry velocity but only marginally by a larger entry mass or a steep [and hence shorter] atmospheric path."[9] They also concluded that, because fragmentation is common even at relatively great heights where the dynamic pressure on the meteoroid is quite small, most meteroids capable of dropping meteorites already possess severe cracks on entry into the atmosphere, a result of previous collisions in space. Halliday and his co-workers also recommended that ground searches be undertaken to find these meteorites (see Chapter 17).[10]

Fig. 16.8. Map of the MORP network showing the location of the 12 stations (filled circles) and the site of the Innisfree meteorite fall relative to some major cities (open stars) and highways in western Canada.

A more extensive study of 259 fireballs observed by MORP by Halliday, Griffin, and Blackwell revealed a significant, but by no means "unambiguous," separation of fireballs into asteroidal and cometary groups.[11] The latter are characterized by a mean entry velocity of 49 km/s, producing brighter, shorter-lived meteors, and the former have a mean entry velocity of 18 km/s and penetrate deeper into the atmosphere. Halliday et al. also estimated a mean density of the cometary group of 0.8 g/cm, near that of water ice; the "asteroidal" group have a median density value of 2.6 g/cm.

We have already noted the operational expensiveness and other shortcomings of traditional meteor observatories like Meanook and Newbrook. We have also seen how the MORP, by combining semi-autonomous operation with a greatly expanded area of coverage, addressed some of these problems. Modern technology has made possible a revolutionary advance in meteor observation. Simple yet sophisticated all-sky cameras can now be set up and operated as meteor observatories by amateur and professional astronomers anywhere. Sandia Labs developed the all-sky cameras and for several years has helped establish camera networks in the United States and Canada. An Alberta

Fig. 16.9. The MORP observatory near Langenburg, Saskatchewan; the other 11 stations are identical to this one. Note the meteor detector mounted on the roof of the building.

Fig. 16.10. The all-sky camera set up on the roof of a building at Athabasca University.

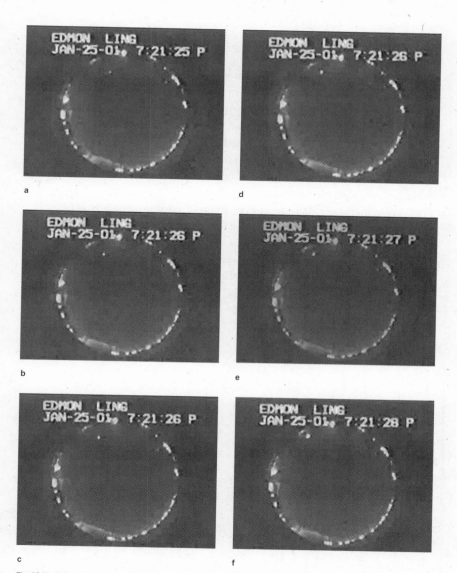

Fig. 16.11. Selected frames from Alister Ling's all-sky camera video of the fireball seen over Edmonton on 25 January 2001. The meteor, already close to the horizon (upper left of images) when it first becomes visible, brightens before disappearing within about three seconds. See http://www.phys.ualberta.ca/video/meteor.html for video footage.

network was instigated by Martin Connors (Athabasca University) with cameras installed at Athabasca University (Fig. 16.10), the University of Alberta, and King's University College. In the Sandia design, a small video camera mounted on a tetrapod views a spherical convex mirror, which produces a fish-eye image of the entire sky. The camera images were originally recorded on VHS tape, which precluded locating the cameras at unattended remote sites. However, the camera signal can also be sent directly to a computer using frame-grabber technology developed for the television broadcast industry. This second method has the added advantage that the digitized images of fireballs can be easily analyzed with the same computer.

All-sky cameras are also operated by amateur astronomers. Alister Ling, a member of the Edmonton Centre of the Royal Astronomical Society of Canada videotaped a bright meteor seen in the southern skies from Edmonton on 25 January 2001 (Fig. 16.11). This fireball was witnessed by many people in central Alberta and may have dropped a meteorite near Red Deer.

17

Stalking the Wild Meteorite

Most of Alberta's meteorites have been found more or less by accident; only Bruderheim, Innisfree, Iron Creek, Peace River, and Vilna were recovered as the result of a deliberate search.[1] The first meteorite search in modern Alberta history was apparently organized by the Edmonton Centre of the Royal Astronomical Society of Canada (RASC) in 1955.[2,3] After receiving a report by Mr. R.E. Cooper of Seba Beach, on Wabamun Lake west of Edmonton, of a suspected meteorite fall, two representatives of the centre went out to interview the farmer. As the result of this investigation a major search, dubbed "Operation WHAM" (short for Wabamun Hunt for Aqua Meteorite), was organized by Franklin Loehde, Ian McLennan, and Earl Milton. Three separate searches of Wabamun Lake were made during July and August using a boat with scuba divers, while ground parties using a mine detector, loaned by the Canadian army, searched the countryside, and a low-flying aircraft searched from above (Fig. 17.1). No meteorite was found, but the operation sustained the enthusiasm of the centre's members, and the experience gained proved invaluable in the search for the Bruderheim meteorite five years later.

Early on the evening of 23 February 1984, a dazzling blue-green and white fireball flew across Alberta in a long flat trajectory from southeast to northwest. Numerous eyewitness accounts and the film from no fewer than seven cameras at four Meteorite Observation and Recovery Project (MORP; Chapter 16) stations established that the fireball may have survived long enough to have exploded and dropped meteorites. Ian Halliday (National Research Council) estimated that the Grande Prairie object had a terminal mass of 12 kg at the explosion, near 30 km height. It is expected that three main fragments with masses in the range of 1–10 kg may have fallen southwest of Grande Prairie.[4] Within several days, a search team from the University of Alberta, the Edmonton Space Sciences Centre (successor to the Queen Elizabeth II Planetarium), and the Grande Prairie Regional College began searching for meteorite fragments but came up empty-handed. A promising report from Ken Smith, an area farmer who said he heard something massive crashing into the trees near his home, was investigated, again without success. A little later, during the Easter weekend, a group

Fig. 17.1 The boat returns to shore with its scuba divers after an unsuccessful search of Wabamun Lake for meteorites.

of RASC Edmonton members and other volunteers mounted their own search of wet, muddy fields and forests, with no success.

Other bright meteors that occasioned newspaper and radio or television coverage, witness interviews, and at least a preliminary search for meteorites include the Frog Lake fireball (16 January 1964); Whitecourt (20 December 1964); Fox Lake (2 March 1985); Jasper (28 September 1985); and Banff (14 October 2001; Fig. 17.2).[5]

From 1985 to 1991, two members of the Edmonton Centre, Peter Brown and Mark Zalcik, investigated several separate MORP fall areas in Alberta to assess their potential for meteorite recovery.[6] For each fall, farmers were interviewed, given information pamphlets, and shown samples of actual meteorites. Brown and Zalcik also evaluated the density of surface rocks at each fall areas (the more rocks present, the harder it is to recognize any meteorites). They also talked to local rock collectors and examined possible meteorite specimens. In addition to revisiting the Grande Prairie site (MORP #925) mentioned above, they investigated Byemoor (#123), Endiang (#171), Lacombe (#683), Edberg (#544), Meeting Creek (#231), Altario (#977), and Fork Lake (#580). A second field campaign was undertaken in the autumn of 1998 and spring of 1999 by Ed Cloutis (University of Winnipeg) and several RASC members. Cloutis et al. searched 10 MORP fall areas including Fork Lake and Edberg in Alberta.

Fig. 17.2. The bright daylight fireball videotaped over Banff National Park on 14 October 2001 by Brad Gledhill.

Although hundreds or thousands of potentially recoverable meteorites have fallen on Alberta since the Manitou Stone, or Iron Creek meteorite, was found by First Nations people (see Chapter 2), we have samples of only 16 of these falls. A biologist concerned with a specific animal or a geologist interested in the Cretaceous–Tertiary boundary layer can usually go somewhere on Earth to collect it, but meteorites can fall anywhere and at any time. Hence, professional meteoriticists depend upon farmers, meteorite collectors, rock hounds, amateur astronomers, and other enthusiasts to find meteorites. As we have seen in preceding chapters, the sooner a meteorite is found after it falls, the better. The scientific community occasionally exhibits an ambivalence about this alliance between collectors and researchers. Some scientists feel that, at worst, every meteorite owned by a collector is one less available for scientific examination and that, at best, collectors drive up the price of meteorites, making it more costly for universities, museums, and other institutions with limited funds to acquire specimens. However, some exemplary private meteorite collectors and dealers provide a valuable service by encouraging meteorite recovery and distribution. In his ambitious effort to promote the scientific study of the Bruderheim meteorite by the widespread distribution of samples, Professor R.E. Folinsbee (University of Alberta) employed the services of Ward's Natural Science Establishment, Inc. (Rochester, New York), a major supplier of geological and biological specimens. As noted in Chapter 8, several research teams obtained their Bruderheim samples from Ward's, helping to make Bruderheim one of the most thoroughly analyzed of all meteorites. Such companies often do basic meteoritical research of their own: New England Meteoritical Services (Mendon, Massachusetts) reported the presence of a huge 22 cm × 14 cm wide clast in the Abee meteorite (Chapter 15).

From time to time, the question of meteorite ownership arises; who owns a meteorite—the person who finds it, the person who owns the land upon which it fell, or some third party, e.g., the government (province, country, etc.)? Lincoln LaPaz (Institute of Meteoritics, Albuquerque, New Mexico; Chapter 6) gave this question considerable thought, taking into account American, French, and even ancient Roman law.[7] LaPaz attacked the prevailing [then as now] legal view that a meteorite belongs to the landowner, saying that the law is prejudicial to the best interests of science and "to public morality." He contended that the law had the effect of either discouraging people from searching for meteorites, knowing they would not be allowed to keep them, prompting finders of meteorites to secretly hoard their finds so as to retain them, thereby denying them to scientific research or causing finders to falsify their accounts of the recovery of the meteorites in order to conceal the true places of find, thereby giving rise to incomplete or untrustworthy data in the scientific literature. Occasionally, there are attempts to enforce the third option, i.e., meteorites belong to the government. In Alberta, shortly after the fall of the Bruderheim meteorite, a member of the Alberta Legislature introduced "a Bill for the Protection of Historical, Archaeological, Paleontological, Ethnic and Meteoric Sites and Objects," stating that any meteorite falling on any land, whether publicly or privately owned, may not be disturbed or removed (likewise any crater, etc. caused by the meteorite) without the authorization of the Minister and, furthermore, every meteorite that falls anywhere in Alberta belongs to the government.[8] Fortunately, this well-intentioned but ill-advised bill never became law. The current Alberta legislation, the *Historical Resources Act*, applies primarily to archaeological and palaeontological objects and sites and makes no mention of meteorites. Canadian (federal) law is explicit: under the *Cultural Property Export and Import Act*, meteorites that fall on private property belong to the property owner and may be held or sold by the owner.[9] However, meteorites are considered *cultural property*, and it is illegal to take or ship them out of Canada without an export permit. The mere act of applying for such a permit alerts interested parties, such as universities and the Geological Survey of Canada, who may then make an offer to the owner to purchase the meteorite. Douglas G. Schmitt, a Vancouver, British Columbia, lawyer interested in meteorites, made a survey of scientists involved in meteorite acquisition in over 20 countries and reported on the relevant laws of each country. Schmitt reported:

> Legal regimes range from a free market to deemed state ownership with no compensation to finders. A free market gives an incentive to searchers but allows ownership by private collectors who do not curate specimens scientifically. Confiscatory laws tempt searchers to conceal or sell finds illegally; or, misrepresent strewn field data to conceal sources, or to make it appear that a meteorite has been legally obtained.[10]

What are meteorites worth? A surprising number of factors can influence a meteorite's value, and several methods have been developed to calculate fair and consistent prices. In 1964, the National Research Council of Canada's Associate Committee on Meteorites (ACOM) developed a pricing formula based on five factors that affect the scientific value of a meteorite.[11] These factors are the type of meteorite; its history, condition, and weight; and whether it is a newly identified fall or a find. Higher ratings are given to rare types such as enstatite or carbonaceous chondrites (10) than ordinary chondrites or octahedrites (1–2). A well-documented specimen from a recent fall is more valuable (6) than a weathered find of unknown or uncertain origin (1). However, the price per gram goes down as total weight recovered goes up.

The ACOM, which was established in 1960 (see Chapter 8) to act as a national clearinghouse for fireball and meteorite data, to co-ordinate and facilitate the study of Canadian meteorites, to promote public awareness of meteorites, and generally to advance the science of meteoritics in Canada, was replaced in 1992 by the Meteorite and Impacts Advisory Committee (MIAC) of the Canadian Space Agency. MIAC was itself superceded in 2007 by two new organizations under a new Canadian Space Agency program: the Astromaterials Working Group and the Small Bodies Working Group. The former has a mandate for promoting the classification of new meteorites and analyzing them with advanced techniques, and the latter has a mandate for tracking fireballs and potential meteorite falls.

18

The Dinosaur Killer

Alberta during the Cretaceous Period was a very different place than it is today. In the west, the Rocky Mountains were being born from the collision between the crust of the Pacific Ocean floor and the crust of the North American continent. To the east, where Saskatchewan is today, a warm sea extended from the Gulf of Mexico to the Arctic Ocean. In between, across the future Alberta, rivers flowed from the mountains down to the sea dropping their loads of rich sediment in broad deltas. On the land, thick vegetation flourished, with lush forests of magnolias, sycamores, figs, ferns, cycads, conifers, and dawn redwood trees covering the slopes and the coastal plains. It was a time when dinosaurs like *Tyrannosaurus* and *Triceratops* still ruled the land.

Then, suddenly, 65 million years ago, 75% of all species of animals and plants on Earth, including all the dinosaurs, inexplicably died. So conspicuous was this Great Dying in the fossil record that geologists chose it to mark the end of the Cretaceous Period and the beginning of the Tertiary Period in geologic time. This boundary between the Cretaceous and Tertiary periods is known as the *K–T boundary* (scientists use the letter K, from the German word for Cretaceous—*Kreide*). The mystery of what killed the dinosaurs as well as more than half the species of life on Earth so long ago has intrigued scientists and the public alike for 200 years. Brilliant detective work on the part of hundreds of scientists in analyzing clues extracted from the fossil record, geological data, and astronomical observations has allowed this mass-extinction mystery to be solved.

The "case-breaking clue" came from an unlikely source. In 1980, the chemist Luis Alvarez and his colleagues Walter Alvarez, Frank Asaro, and Helen Michel (University of California, Berkeley) announced the discovery of a thousand times more iridium than is usual in the Earth's crust in the K–T boundary at Gubbio, Italy, where it is represented by a thin layer of clay.[1] Alvarez and his colleagues realized that something extraordinary must have happened to deposit this much excess iridium—the so-called *iridium anomaly*. They hypothesized that an asteroid struck the Earth 65 Ma ago, and the resulting explosion blanketed the world with a layer of dust. The study of meteorites, asteroids, and comets has shown that these objects contain a higher proportion of iridium than the terrestrial average; this would explain the excess iridium in the K–T boundary layer. They further hypothesized that it was the environmental consequences of this asteroid impact that caused the K–T mass extinction.

Scientists believe that a huge asteroid or comet, estimated to be 15–30 km in diameter, struck the Earth. The impact triggered stupendous tsunamis that scoured the world's coasts. The entire globe shuddered from a massive earthquake. The heat from the fireball and the fiery re-entry of the ejected debris scorched the face of the planet; creatures everywhere were literally broiled to death; vegetation burst into flames; continent-sized forest fires raged. The impactor was pulverized and its material joined the trillions of tons of vaporized rocks in the dust-laden atmosphere. During the next few years, sulphur in the dust and nitrous oxides created by the heat of impact combined with water to make acid rain that corroded the land and poisoned the oceans. The massive injection of energy into the atmosphere spawned hurricanes and storms of unimaginable fury. After the fire, the acid rain, the earthquake, and the tsunami, a vast cloud of soot and dust enshrouded the Earth, cutting off all sunlight everywhere. Complete darkness descended upon the world for half a year. Plants died from lack of photosynthesis, and animals starved or froze to death. As the dust and soot settled out of the atmosphere and the air cleared, lethal ultraviolet radiation, with no ozone to block it, seared the planet.[2]

Support for the Alvarez hypothesis came quickly. Within just a few years, the K–T boundary layer with its characteristic iridium anomaly had been found in many places around the world, including Alberta.

The extreme conditions of heat (several thousand degrees Celsius) and pressure (100 gigaPascals, which is a million times normal atmospheric pressure of ~100 kPa) created during a meteorite or comet impact are far in excess of those found in volcanic eruptions. These conditions produce a distinctive change in the rocks at the impact site called *shock-metamorphism*. Hence, the presence of shocked minerals is conclusive evidence of an impact. In 1984, another link between an impact and the cause of the K–T extinction was forged by the discovery of *shocked quartz* grains, recognizable by the grooves on their surface, in the K–T layer.

If an object big enough to cause a global environmental catastrophe had really struck the Earth, where was the crater? After many known impact craters were investigated then rejected as being too large, too small, too old, or not old enough, Alan Hildebrand, then at the Geological Survey of Canada in Ottawa, drew attention to the 180 km wide Chicxulub crater buried under the Yucatan Peninsula in Mexico.[3] Here, at last, was "the scene of the crime"—a crater of just the right size and age. Now all that was needed was a "re-enactment of the crime": In July 1994, a comet, Shoemaker-Levy 9, previously shattered to pieces by a close encounter with Jupiter, slammed into that planet just as our purported comet struck the Earth 65 million years ago.[4] Only, this time round, there were astronomers with telescopes and spacecraft present to observe and record the event. As the individual comet fragments, none larger than an estimated 1 km in diameter, hurtled into Jupiter's atmosphere they produced titanic fireballs that, even from 800×10^6 km away, blinded the sensitive instruments on telescopes. Dust from the disintegrated comet chunks formed dark splotches on Jupiter larger than the entire Earth—had one of these comparatively small objects hit our planet instead, the result would have been devastating.

The original Gubbio boundary layer samples studied by Alvarez et al. and most of the subsequently discovered world-wide occurrences were laid down in ancient *marine* environments, but the Alberta locality is of particular interest because it was formed on *land*. In Alberta, the K–T boundary layer can be seen near Coal Valley in the Rocky Mountain foothills and in the Badlands east of Huxley.[5] Here, the Red Deer River has cut a canyon hundreds of metres deep through the rocks, slicing through the Cretaceous–Tertiary boundary and, thus, exposing the section we are interested in. The K–T boundary layer occurs as a thin (2–3 cm) layer of soft, reddish clay immediately below a ~0.5 m thick coal seam.

David Carlisle in his splendid account of his hunt for evidence of the Cretaceous–Tertiary boundary event in the Badlands might well have been describing my own adventures with my 10-year-old daughter Tanya.[6] Having previously done a science fair project on the meteorite craters of North America, Tanya decided to search for the K–T boundary layer in the valley of the Red Deer River for her next project. Having obtained permission from the farmer and his wife on whose land the layer was exposed, we gingerly clambered down the steep clay embankment (if it had rained during our visit, the cliff would have been too slippery to climb back up) and, with the aid of maps and diagrams from scientific reports, located the boundary layer (Fig. 18.1).

The actual boundary layer proved hard to discern, the strata often blending imperceptibly into each other, so Tanya also took several samples of clay from above and below the probable K–T boundary. Back home, she air-dried, crushed, and sieved the samples; then following the method described by R.A.F. Grieve and J. Alexopoulis (Geological Survey of Canada, Ottawa) she separated out the quartz and other heavy grains.[7] She examined the grains under a microscope, estimating there were about 4500 large and medium-sized grains per microscope slide of which about one-third were clear (likely quartz and feldspar), one-half were amber-coloured (probably mica), and the rest

Fig. 18.1. Tanya Whyte and the Cretaceous–Tertiary boundary. The layer is between the black coal seam and the light-coloured sandstone. Stripes on the stick measure 10 cm each.

were dark material (small clumps of clay particles, microtektites, and soot). Of roughly 27,000 clear grains examined, Tanya found four definite shocked quartz grains (Fig. 18.2).

The 180 km diameter Chicxulub crater identified by Hildebrand as the "dinosaur killer" is completely buried 1 km deep under more recent rocks, and there are few visible signs of its existence on the surface.[8] It was only discovered by analysis of gravity and magnetic field data, supplemented by examination of drill core samples, mostly collected in the course of petroleum exploration. As of 2005, about 175 impact craters have been positively identified on Earth; they are also called astroblemes ("star scars") or impact structures. Many provinces and territories in Canada have visible variously sized impact craters, but Alberta has only two confirmed impact structures, both buried, and a few tentatively identified ones.[9]

Alberta is one of the most thoroughly geophysically explored regions in the world, again as the result of oil and gas exploration. Over the past 50 years, large amounts of seismic data have been acquired within Alberta. Some of these data show enigmatic circular features that are not easily explained as ancient reefs, diatremes (breccia-filled volcanic pipes that are also of great interest in Alberta as possible sources of diamonds), or other round geological structures. These

Fig. 18.2. A long clear grain (arrowed) of shocked quartz from the K–T boundary layer. Note planar features (crossed grooves) near black wavy line; the latter is an artifact from the slide mounting fluid.

buried circular features show many characteristics diagnostic of impact craters such as raised rims, annular synforms (ring-shaped depressions or "moats"), raised central uplifts, and breccia infill (material that has slumped into the crater), particularly when imaged with three-dimensional computer modelling software. The 25 km wide Steen River structure, located in northwestern Alberta at 59°30′N, 117°38′W, is a classic example of a complex impact crater whose identity has been confirmed by the presence of shocked minerals in core samples.[10] It has an estimated age of 95 million years and is buried under 200 m of Cretaceous and recent sediments that have protected it from erosion.

The Eagle Butte structure, variously reported as 10 or 15 km in diameter, is located at the west end of the Cypress Hills at 49°42′N, 110°30′W in southeastern Alberta. Unlike the Steen River structure, which is too deeply buried to reveal any surface evidence of its existence, a faint circular outline of Eagle Butte can be seen on aerial photographs of the region (Fig. 18.3). The anomalous nature of Eagle Butte was noted from seismic and drill core data as long ago as 1930.[11] However, H.B. Sawatzky, a Calgary geophysicist (Francana Oil & Gas Limited) is generally credited with drawing attention to its probable impact origin.[12] In seismic scans, Eagle Butte displays a prominent central uplift, an annular synform, and raised rim—characteristics of a complex impact crater.[13]

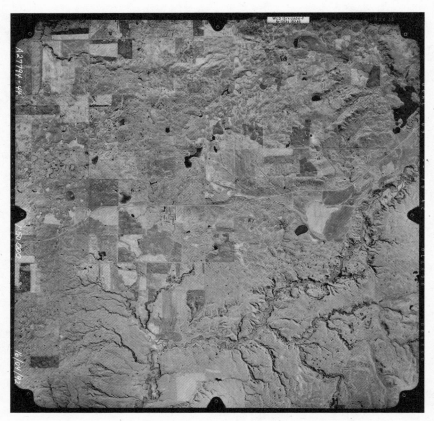

Fig. 18.3. The faint, vaguely circular outline of the Eagle Butte impact structure can just be made out near the centre of this aerial photograph of the region near the western end of the Cypress Hills, south of Medicine Hat.

Its identity as an impact crater was clinched with the discovery by D.C. Lawton, R.R. Stewart (both of the University of Calgary) and R. Gault of shatter cones at the surface (Fig. 18.4).[14]

The Hotchkiss structure, located near the foothills of northwestern Alberta, is an enigmatic feature whose appearance based on seismic data closely resembles that of a complex impact crater.[15] Hotchkiss has a central uplift region, large scale faulting at the rim, a breccia infill in the central uplift, and a continuation of disturbance in the underlaying rocks to more than 1500 m below the top of the feature. It is estimated to have been 4.5 km in diameter and 500 m deep at the time it was formed, between 120 and 330 Ma ago. It is now buried under approximately 1 km below the surface, centred on 57°32'13"N,

Fig. 18.4. Shatter cone from Eagle Butte found by Mark Zalcik. Scale is in centimetres.

118°52′42″W. [16] Geologists Michael J. Mazur and Robert R. Stewart (University of Calgary) reconstructed the events leading to the the formation of the Hotchkiss structure as it appears today:

> If the impactor is assumed to have been an iron object traveling at 20 km/s, we find that the diameter required to create a 4.5 km crater is about 240 m. As the iron meteoroid crashed into what was likely a shallow sea, the excavation of the transient cavity and start of the central uplift began. Within about 20 seconds, the central uplift was nearly formed and the rim was beginning to collapse. After several minutes, the crater had assumed its final shape. A period of erosion followed the formation of this structure whereby nearly 200 Ma of the sedimentary record in this area was erased. [17]

The James River structure located in south-central Alberta is another possible complex impact crater 4.8 km wide. It has an annular synform and a central uplift 2.4 km in diameter (Fig. 18.5). [18] It is buried under nearly 5 km of sediments and is estimated to have been formed sometime between Middle Devonian to Late Cambrian time, i.e., about 400–500 Ma ago.

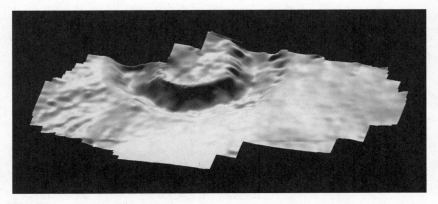

Fig. 18.5. A computer model of the James River impact structure created from three-dimensional seismic data. Note the annular moat and central uplift.

In 1997, Hans-Henrik Westbroek, then a University of Calgary graduate student, proposed in his master's thesis that the bowl-shaped Purple Springs structure buried in south-central Alberta at 49°52′30″N, 111°52′00″W is an impact crater. [19] Westbroek used drill core data from four wells straddling Purple Springs and data from three seismic lines that pass completely over it to delineate its structure. Purple Springs appears slightly elliptical in plan view with a major axis diameter of about 4 km and a minor axis diameter of about 3 km. In these features, Purple Springs resembles a simple impact crater, but a typical simple crater has a depth:diameter ratio of about 1:5, whereas the Purple Springs structure is much shallower, with a ratio of about 1:20. Westbroek noted the faulting and terracing seen along the rim of Purple Springs are more characteristic of a complex impact crater, but there is no evidence of a structurally uplifted central peak—another charactersitc of complex craters. He concluded that Purple Springs may be a transitional simple-complex impact crater that has been heavily eroded, obscuring its true nature.

Meteorites continue to rain down on the Earth at an estimated rate of almost 14,000 falls larger than 100 g per year, although normally fewer than a dozen of these are recovered.[1] When I began writing this book there were 14 known Alberta meteorites; then the provisionally named Belly River Buttes meteorite (see Chapter 7) was identified in 2004, although it had originally been found in 1992. Since then, yet another Alberta meteorite has been identified. The also provisionally named Lake Eliza meteorite was identified in 2005.

The Lake Eliza meteorite was found by Alan Zalaski, a junior high science and physical education teacher at Glen Avon School in St. Paul, about 200 km northeast of Edmonton.[2] A rock collector since childhood, Zalaski has, for years, set aside rocks that he suspected might be meteorites. For the last several years, he and his friend and neighbour, Robert Tymofichuk, and their children have been exploring the farmland surrounding Lake Eliza, looking for old bison skulls, antlers and unusual rocks. It was on one of these excursions, probably during the summer of 2003, that Zalaski found a rectangular, reddish-brown rock that he added to his collection of possible meteorites. By that time, he was a more sophisicated collector, having read many books on meteorites and become familiar with the characteristics of meteorites, e.g., presence of a fusion crust, heavier-than-usual, magnetic, etc. He contacted Chris Herd, Curator of the University of Alberta Meteorite Collection, to discuss his specimens and was about to take the rocks to Herd for examination when he learned that Tom Weedmark, a representative of the Prairie Meteorite Search team (see below), would be visiting St. Paul in August 2005. Zalaski and Robert Tymofichuk took a dozen of their likeliest meteorite candidates to the public meeting at the library in St. Paul to show to Weedmark. The latter found six of the rocks "interesting" and thought that one (Fig. 19.1) was definitely a meteorite.

Two years after finding the rock identified by Tom Weedmark as a meteorite, Zalaski and Tymofichuk can no longer remember exactly where they found it. The best they can do is to narrow down the location to somewhere within the sections of land bordering Lake Eliza to the north, east, and south.[3]

Fig. 19.1. The Lake Eliza meteorite found by Alan Zalaski. The small round bumps protruding from the surface are chondrules, which are more resistant to weathering than the matrix surrounding them.

Zalaski loaned the 340 g meteorite to the Prairie Meteorite Search; in turn, they took it to Alan Hildebrand at the Department of Geology and Geophysics at the University of Calgary. From the degree of weathering, Hildebrand estimated that the meteorite fell to Earth several hundred years ago. Analysis of a small section sawed off the meteorite (Fig. 19.2) showed it to be an H chondrite. The metal (nickel-iron) content of Lake Eliza is so high that the sawn face of the meteorite glints brilliantly in sunlight (Fig. 19.3). Rust-filled cracks, also visible on the sawn face, are reminiscent of those in Belly River (see Fig. 6.3) and indicative of weathering.

The Prairie Meteorite Search was initiated by Alan Hildebrand (Department of Geology and Geophysics, University of Calgary), Peter Brown (Department of Physics and Astronomy, University of Western Ontario), and Martin Beech (Department of Physics, University of Regina) to promote the recovery of meteorites in Canada. The Prairie Meteorite Search uses the method pioneered by the American meteorite collector Harvey Nininger (see Chapter 3) 70 years ago in the southern and midwestern states to recover meteorites. The method involves using publicity campaigns, mainly through putting articles in local newspapers, to alert residents of a rural area that they can bring in any suspected meteorites to a public meeting at the local school, library, etc. As an added inducement, there is a display of actual meteorites, so that everyone may see what "the real thing" looks like.

Fig. 19.2. Kelsea Zalaski holding the Lake Eliza meteorite. Note where a sample section was sawed off (left).

Still other meteorites continue to be recovered in Alberta. Since the recovery of the Lake Eliza meteorite in 2005, two more meteorites have been identified by Chris Herd (Curator of the University of Alberta Meteorite Collection) and Andrew Locock (Collections Administrator, Department of Earth and Atmospheric Sciences, University of Alberta). Herd, Locock, and others are currently working to classify these meteorites and have them officially named and recognized by the Meteoritical Society.[4]

Alberta meteorites continue to be studied by researchers. Abee, Belly River, and Bruderheim were included in a 2003 survey paper on the porosities and densities of stony meteorites by D.T. Britt (University of Central Florida) and G.J. Consolmagno (University of Arizona).[5] Daniel Garrison, Sara Hamlin, and Donald Bogard (Johnson Space Center, Houston) measured the chlorine abundance in 94 meteorites, including Abee and Bruderheim, using neutron activation analysis.[6] In his geochemical and petrographic study of some minor and rare minerals in several meteorites, Weibiao Hsu (Washington University in St.

Fig. 19.3. Close-up of the sawn face of Lake Eliza; mirror-like flakes of nickel-iron alloy glint in sunlight, revealing the high metal content of this H chondrite. Note also the rust-filled cracks.

Louis) reported on rare earth element concentrations in albite and the composition of oldhamite in Abee.[7] Without mentioning Abee (or any other meteorite) specifically, Stanley Love and Klaus Keil (University of Hawaii, Manoa) presented a detailed analysis of the possibility of Mercurian meteorites being launched from that planet to Earth and the feasibility of identifying such meteorites. They concluded that the delivery of Mercurian rocks to Earth is feasible, but the efficiency of the process would be <1% of that computed for Mars. They also suggested that recognizing Mercurian meteorites would be difficult without knowing more about the elemental and minealogical composition of the planet.[8] A team of Japanese researchers at Kyushu University and the University of Tokyo measured the cosmic-ray exposure ages of nine enstatite meteorites from Antarctica and two non-Antarctica ones. Combining their data with published data for Abee and several other enstatite chondrites, they concluded that there is no correlation between the cosmic-ray exposure ages of EH and EL chondrites and their petrologic types and chemical classification.[9]

In 2001, Andrea Patzer and Ludolf Schultz (Max-Planck-Institut für Chemie, Mainz, Germany) reported new measurements of cosmogenic and radiogenic noble gas—helium (He), neon (Ne), argion (Ar), and xenon (Xe)—concentrations, cosmic-ray exposure ages, and gas retention ages for 57 enstatite chondrites, including Abee.[10] Like the Japanese study (above), they also found no apparent systematic trend in the distribution of cosmic-ray exposure ages of EH and EL chondrites. They reported that several enstatite chondrites yielded significantly lower ^{38}Ar ages compared to those calculated from cosmogenic ^{3}He and ^{21}Ne. They suggested that these discrepancies might be the result of the loss of ^{38}Ar from terrestrial weathering on Earth and conducted experiments that confirmed this hypothesis. In a subsequent paper, Patzer and Schultz reported on the trapped noble gases in enstatite chondrites.[11] Trapped gases are those, like presolar gases, that were incorporated into the meteorite when its parent body was first created and were not produced *in situ* by radioactive decay, for example.

Bruderheim was recently used in an important study by R.H. Brazzle and his colleagues (Washington University in St. Louis) that confirmed the reliability of the iodine-xenon (I-Xe) chronometer, a dating method based upon the radioactive decay of ^{129}I into ^{129}Xe.[12]

Innisfree was also the subject of a 2005 paper on the fragmentation of meteorties during atmospheric flight by Z. Ceplecha and D.O. Revelle.[13] A re-examination of some 50 shocked ordinary chondrites, including Innisfree, by Alan Rubin (University of California, Los Angeles) in 2004 suggested that their thermal and shock histories may be more complex than previously thought.[14] Rubin found evidence that some chondrites have been shocked, then annealed by heat, then reshocked; some may have experienced multiple episodes of shock and annealing. R.J. Reisener and J.I. Goldstein (University of Massachusetts), in their second paper on the metallography of ordinary chondrites, pointed out that shock-generated Neumann lines (see Chapter 2) are commonly present in the kamacite particles of ordinary chondrites. They reported that, in Innisfree, the Neumann lines are curved, suggesting that plastic deformation followed the shock process.[15]

The existence of thoroughly investigated meteorites like Abee and Bruderheim is very helpful in the study of Antarctica and North African meteorites now being found in large numbers, as well as rare and exotic meteorites from Mars. The meteorites of Alberta will continue to be studied as samples are ever more widely distributed and new analytical methods become available, adding to the growing body of scientific literature.

Appendix

Periodic Table of the Elements

1.00794 1 **H** Hydrogen								
6.941 3 **Li** Lithium	9.012182 4 **Be** Beryllium							
22.989768 11 **Na** Sodium	24.3050 12 **Mg** Magnesium							
39.0983 19 **K** Potassium	40.078 20 **Ca** Calcium	44.955910 21 **Sc** Scandium	47.88 22 **Ti** Titanium	50.9414 23 **V** Vanadium	51.9961 24 **Cr** Chromium	54.93805 25 **Mn** Manganese	55.847 26 **Fe** Iron	58.93320 27 **Co** Cobalt
85.4678 37 **Rb** Rubidium	87.62 38 **Sr** Strontium	88.90585 39 **Y** Yttrium	91.224 40 **Zr** Zirconium	92.9064 41 **Nb** Niobium	95.94 42 **Mo** Molybdenum	98 43 **Tc** Technetium	101.07 44 **Ru** Ruthenium	102.9055 45 **Rh** Rhodium
132.9054 55 **Cs** Cesium	137.327 56 **Ba** Barium	57–71 **La-Lu**	178.49 72 **Hf** Hafnium	180.9479 73 **Ta** Tantalum	183.85 74 **W** Tungsten	186.207 75 **Re** Rhenium	190.23 76 **Os** Osmium	192.22 77 **Ir** Iridium
223.0197 87 **Fr** Francium	226.0254 88 **Ra** Radium	89-103 **Ac-Lr**	261.1138 104 **Rf** Rutherfordium	262.1138 105 **Db** Dubnium	263.1182 106 **Sg** Seaborgium	262.1229 107 **Bh** Bohrium	265 108 **Hs** Hassium	266 109 **Mt** Meitnerium

Atomic Weight

24.3050 12 **Mg** — Symbol Magnesium — Element

Atomic Number

138.9055 57 **La** Lanthanum	140.115 58 **Ce** Cerium	140.90765 59 **Pr** Praseodymium	144.24 60 **Nd** Neodymium	146.9151 61 **Pm** Promethium	150.36 62 **Sm** Samarium	151.965 63 **Eu** Europium
227.0278 89 **Ac** Actinium	232.0381 90 **Th** Thorium	231.0359 91 **Pa** Proactinium	238.03 92 **U** Uranium	237.0482 93 **Np** Neptunium	244.0642 94 **Pu** Plutonium	243.0614 95 **Am** Americium

						4.002602 2
						He
						Helium

10.811 5	12.011 6	14.00674 7	15.9994 8	18.9984032 9	20.1797 10
B	**C**	**N**	**O**	**F**	**Ne**
Boron	Carbon	Nitrogen	Oxygen	Fluorine	Neon
26.981539 13	28.0855 14	30.973762 15	32.066 16	35.4527 17	39.948 18
Al	**Si**	**P**	**S**	**Cl**	**Ar**
Aluminum	Silicon	Phosphorus	Sulphur	Chlorine	Argon

58.6934 28	63.546 29	65.38 30	69.723 31	72.61 32	74.92159 33	78.96 34	79.904 35	83.80 36
Ni	**Cu**	**Zn**	**Ga**	**Ge**	**As**	**Se**	**Br**	**Kr**
Nickel	Copper	Zinc	Gallium	Germanium	Arsenic	Selenium	Bromine	Krypton
106.42 46	107.868 47	112.411 48	114.818 49	118.710 50	121.75 51	127.60 52	126.9045 53	131.29 54
Pd	**Ag**	**Cd**	**In**	**Sn**	**Sb**	**Te**	**I**	**Xe**
Palladium	Silver	Cadmium	Indium	Tin	Antimony	Tellurium	Iodine	Xenon
195.08 78	196.96654 79	200.59 80	204.3833 81	207.2 82	208.98037 83	208.9824 84	209.9871 85	222.0176 86
Pt	**Au**	**Hg**	**Tl**	**Pb**	**Bi**	**Po**	**At**	**Rn**
Platinum	Gold	Mercury	Thallium	Lead	Bismuth	Polonium	Astatine	Radon
269 110	272 111	277 112						
Uun	**Uuu**	**Uub**						
Ununnilium	Unununium	Ununbium						

157.25 64	158.92534 65	162.50 66	164.93032 67	167.26 68	168.9342 69	173.04 70	174.967 71
Gd	**Tb**	**Dy**	**Ho**	**Er**	**Tm**	**Yb**	**Lu**
Gadolinium	Terbium	Dysprosium	Holmium	Erbium	Thulium	Ytterbium	Lutetium
247.0703 96	247.0703 97	251.0796 98	252.0829 99	257.0951 100	258.0986 101	259.1009 102	262.11 103
Cm	**Bk**	**Cf**	**Es**	**Fm**	**Md**	**No**	**Lr**
Curium	Berkelium	Californium	Einsteinium	Fermium	Mendelevium	Nobelium	Lawrencium

Introduction

1. In recent usage, the term "meteoroid" has come to be defined as a natural space object having a size between 100 μm and 10 m, although some researchers add the qualification that the object must enter the atmosphere to be called a meteoroid. See Beech and Steel (1995) for a discussion of this subject.
2. Although 6% is a modest nickel content for an iron meteorite, it would be a fabulous level for nickel ore on Earth—the rich "Ovoid" at the Voisey's Bay deposit in Labrador has an average grade of 2.83% Ni, and that is considered very rich.
3. Rose (1863).
4. LaPaz and LaPaz (1961), p. 132.
5. Mason (1966).
6. Extract from a letter to M. Luski from C. Herd (5 September 2007). Herd says: "What is confusing is that bronzite in terrestrial rocks is orthopyroxene with between 10 and 30% of the iron-rich end-member (ferrosilite), and hypersthene is orthopyroxene with between 30 and 50% ferrosilite. In contrast, in the meteorite classification scheme, bronzite refers to orthopyroxene with 10 to 20% ferrosilite and hypersthene to orthopyroxene with 20 to 30% ferrosilite."
7. Andrew Locock, email message to author, 20 March 2007. Used with permission.
8. For example, Schultz and Kruse (1989) and Mason (1971).
9. Melcher (1981).

Edmonton (Canada)   1

1. Royal Astronomical Society of Canada (1939), p. 270.
2. Extract from a letter to J.A. Allan from P.M. Millman (31 May 1939). This and subsequent extracts from Allan's correspondence are used with permission from the Department of Earth and Atmospheric Sciences, University of Alberta, Edmonton.
3. Extract from a letter to P.M. Millman from J.A. Allan (15 June 1939).
4. Nininger (1940), p. 557.
5. Extract from a letter to J.A. Allan from F.C. Leonard (26 September 1939). Leonard, incidently, was the founding father of the Meteoritical Society; see Clarke and Plotkin (2002).
6. Extract from a letter to F.C. Leonard from J.A. Allan (26 October 1939).
7. Allan (1953).
8. Allan (1953).
9. J.A. Kelso, Report of Analysis. University of Alberta, 19 June 1939.

10. See also Wasson (1974), pp. 160–161; Dodd (1981), p. 196; and Buchwald (1975), p. 554 for additional photographs of the Neumann lines in the Edmonton meteorite.
11. The astute reader will have realized that a cube is a six-sided solid, or *hexahedron*, hence the name hexahedrite.
12. Buchwald (1975).
13. Buchwald (1975), pp. 554–555.
14. Testing the hardness of metals and other materials, by measuring the indentation of a sample by a diamond-tipped anvil or probe applied with a known force, is a standard metallurgical procedure. The *microhardness* of a meteorite or other sample can be determined with great accuracy by using a very small diamond anvil and measuring the indentation under high magnification with a precision microscope.
15. Folinsbee, R.E. (1953). This five-page report is undated and gives no author, but I have attributed it to Folinsbee on the assumption that it is the "detailed information about the Edmonton meteorite" referred to in a letter to R.E. Folinsbee from P.M. Millman (1953).
16. Extract from a letter to R.E. Folinsbee from P.M. Millman (13 March 1953).
17. Extract from a letter to R.E. Folinsbee from R. Kellogg (12 March 1956).
18. Bauer (1963).
19. Lovering et al. (1957).
20. Scott and Wasson (1975), p. 542.
21. Extract from a letter to R.E. Folinsbee from J.T. Wasson (28 April 1966).
22. Wasson (1969), p. 861.
23. Scott and Wasson (1975), p. 527.
24. Extract from a letter to R.E. Folinsbee from H. Voshage (28 April 1969).
25. Pernicka et al. (1979).
26. Buchwald (1975), Vol. 1. p. 555.
27. Weinke (1979).

Iron Creek

2

1. MacGregor (1954). Permission to quote extensively provided by the late J.G. MacGregor's son.
2. Henry and Thompson (1810), p. 622.
3. Earl of Southesk (1875), p. 423.
4. Cheadle (1931), p. 127.
5. Viscount Milton and Cheadle (1865), pp. 169–170.
6. Milton and Cheadle (1865), p. 169.
7. Spratt (1989), p. 84.
8. McDougall (1902), pp. 141–142.
9. Coleman (1886).
10. MacGregor (1954), p. 251.
11. Allen Ronaghan was kind enough to send me a copy of this undated and unsigned document, titled "Iron Creek (Victoria)," which was found at his instigation in the files of the Royal Ontario Museum. He says, in a footnote in his paper (Ronaghan, 1973), that it was obviously prepared by someone who knew the circumstances of the moving of the meteorite. Courtesy of the Royal Ontario Museum.

12. Millman and McKinley (1967), p. 278.
13. Butler (1872), pp. 304–305.
14. Millman (1953b), p. 162.
15. MacGregor (1954), p. 251.
16. Ronaghan (1973), p. 10.
17. Allen Ronaghan says that this so annoyed Alberta visitors [the author among them] that the Royal Ontario Museum eventually changed the label to "Iron Creek, Alberta."
18. Ronaghan (1973), p. 12.
19. See note 11 above.
20. Millman (1953a), p. 30.
21. Hey (1966), p. 215.
22. Ronaghan (1973), p. 12.
23. Extract from a letter to R.E. Folinsbee from F.J. Howell (1988).
24. Lowey (1993).
25. R.E. Folinsbee, conversation with Bill McDonald, Edmonton, 15 January 1964.
26. R.E. Folinsbee, field notes from excursion to Hardisty, 24–26 September 1971.
27. Extract from a letter to J. MacNeill from R.E. Folinsbee (1991).
28. Coleman (1886).
29. Farrington (1907). See plate 35.
30. Farrington (1907), p. 115.
31. Krinov (1960b), p. 196. Krinov notes that this pattern was first observed by the Englishman Thompson in 1804. However, Widmanstätten published or, at least, publicized his discovery first.
32. This implies a total cooling time of ~100 Ma to 1 Ga; thus, the asteroid's iron core must have been shrouded in a thick mantle or crust of insulating rock.
33. Lovering et al. (1957), p. 277.
34. Scott et al. (1973), p. 1961.
35. Haack and Scott (1993), p. 3458.
36. Malvin et al. (1984), p. 804.
37. Spratt (1989), p. 89.
38. Coleman (1886), p. 97.
39. After an extended lobbying effort by various individuals and community groups in Alberta and negotiations by Ralph Sorenson, MLA for the Iron Creek area constituency, and Horst Schmid, Alberta's Minister of Youth, Culture and Recreation, the Royal Ontario Museum returned the Iron Creek meteorite to Alberta in 1973, "on permanent loan." In December 2001, Victoria University (Toronto) transferred ownership of the meteorite to the Provincial Museum of Alberta (now the Royal Alberta Museum).
40. One of the reviewers of this manuscript pointed out that the revised mass for Iron Creek "makes the Madoc iron from Ontario (the first meteorite to be acquired by the fledgling Geological Survey of Canada in the 1850s) the heaviest single meteorite mass recovered in Canada; this still leaves Alberta with 3 of the 4 heaviest total recoveries (Bruderheim, Madoc, Iron Creek, and Abee)."
41. Spratt (1989), p. 89.
42. Ron Mussieux, email to author, 16 September 2005.
43. Blakeslee (2002), pp. 52–59.

Kinsella 3

1. Millman (1953a), p.30.
2. Scott et al. (1977), p. 425.
3. Scott et al. (1978), pp. 339–340.
4. Scott et al. (1977), p. 432.
5. Scott et al. (1977), p. 432.
6. Kracher et al. (1980), p. 774.
7. Malvin et al. (1984), pp. 803–804. They wrote, "We now claim treasure hunter's rights for these data." Treasure hunting in this context means sifting through old scientific observations to uncover discarded, unnoticed, or under-utilized data.
8. Haack and Scott (1993), p. 3458.
9. Chris Herd, Curator of the University of Alberta Meteorite Collection, told the author that UCLA is presently not interested in selling or trading either the main mass or a sample of Kinsella.

Mayerthorpe 4

1. Extract from an undated [circa September 1964] memo, titled "Mayerthorpe Meteorite," recording details of the purchase of the Mayerthorpe [#1] meteorite. This and subsequent extracts from R.E. Folinsbee's correspondence are used with permission from the Department of Earth and Atmospheric Sciences, University of Alberta.
2. Folinsbee (1964). From a copy of the Mayerthorpe report Folinsbee sent to E.L. Krinov, editor of the *Meteoritical Bulletin*; the number and date of this *Bulletin* were provided in Buchwald (1975).
3. Extract from a bill of sale [for Mayerthorpe #1] from the University of Alberta Department of Geology, signed by Mike Dmitroca and witnessed by William Dmitroca on 19 September 1964.
4. Extract from a bill of sale [for Mayerthorpe #2] from the University of Alberta Department of Geology, signed by Walter Dmitroca and witnessed by R.E. Folinsbee on 6 October 1964.
5. Extract from an undated and unsigned memo, entitled "Mayerthorpe #2."
6. Folinsbee (1965). Extract from a letter to E.L. Krinov from R.E. Folinsbee (7 October 1964). According to Buchwald (1975), this information was published in *Meteoritical Bulletin* No. 33.
7. Extract from a letter to R.E. Folinsbee from H. Voshage (7 June 1968).
8. Extract from a letter to H. Voshage from R.E. Folinsbee (18 June 1968).
9. Extract from a letter to R.E. Folinsbee from H. Voshage (16 January 1969).
10. Voshage (1978).
11. Voshage (1981). One early reviewer of the manuscript has suggested that this ratio should be about 12.0, not 1.2, and that, like many early attempts at Li isotope ratio measurements, it is probably unreliable.
12. Aylmer et al. (1988).
13. Wasson (1970).
14. Choi et al. (1995), p. 596.
15. Wasson and Kallemeyn (2002).

Millarville

5

1. Extract from a letter to J.T. Wasson from someone (unknown) writing at the behest of E. Klovan to provide details about the finding of the Millarville meteorite (29 November 1977).
2. This corresponds to latitude 50°47′50″N, longitude 114°18′34″W.
3. The presentation of scrolls to discoverers of first specimens of Canadian meteorites was approved in principle at ACOM's 20th meeting in November 1976; the first such scroll was presented to Mr. C.E. Hayward.
4. University of Calgary Gazette (2001).
5. Harvey (1979).
6. Kracher et al. (1980), p. 774.
7. Wasson and Richardson (2001), p. 952.

Belly River

1. LaPaz (1953), p. 106.
2. After graduation, Wettlaufer did field work at the Morlach and Long Creek sites in Saskatchewan and Head-Smashed-In Buffalo Jump in Alberta. He could not find a permanent job in archaeology in Canada (he told the author "they were hiring Americans and Britishers, but not Canadians"), so he took a job as an archaeologist in New Mexico. His work at Head-Smashed-In Buffalo Jump would lead eventually to its designation as a UNESCO World Heritage Site in 1981 and to its development as a world-class museum and interpretative centre. In January 2004, the Governor General of Canada announced the appointment of Boyd Nickolas David Wettlaufer to the Order of Canada for "his groundbreaking archaeological work in western Canada," no mention being made of his meteorite discovery. At the time of writing, the 92-year-old Wettlaufer lived in Victoria, BC.
3. B. Wettlaufer, telephone conversation with author, 11 March 2004.
4. LaPaz (1953).
5. Boyd Wettlaufer told the author he did not see LaPaz's paper and become aware of the mistaken information until the 1970s. Not wishing to cause his old friend any trouble, he said nothing at this time. Lincoln LaPaz died of natural causes at age 88 on 19 October 1985. It was only when I contacted Wettlaufer in early 2004 that he decided to set the record straight.
6. Canada NTS maps, 1:50,000 scale, sheets 82 H/12 (Brocket) and 82 H/11 (Fort Macleod). Wettlaufer's retraced path follows the right (eastern) edge of map 82 H/12 and ends on the left (western) edge of map 82 H/11. LaPaz's published location is down in the right bottom (southeast) corner of map 82 H/11.
7. Wettlaufer and Whyte (2004).
8. Wilson (2004), p. ii.
9. LaPaz (1953), p. 106.
10. In UCLA meteoriticist Frederick Leonard's notation—not to be confused with CV, a carbonaceous chondrite, in more recent classification schemes.
11. LaPaz (1953), p. 108.

12. Extract from a letter to R.E. Folinsbee from L. LaPaz (1960). This and subsequent extracts from Folinsbee's correspondence are used with permission from the Department of Earth and Atmospheric Sciences, University of Alberta.
13. Extract from a letter to R.E. Folinsbee from L. LaPaz (1961).
14. Extract from a letter to R.E. Folinsbee from K.R. Dawson (1961).
15. Mason and Wiik (1967).
16. Urey and Craig (1953).
17. Mason (1963), p. 1014.
18. Stöffler et al. (1991).
19. Stöffler et al. (1991), p. 3859.
20. Wilson (2004), pp. 4–5.

Belly River Buttes 7

1. The geographical location is latitude 49°33.712′N, longitude 113°15.07′W, NAD27. Alan Hildebrand, email message to author, 15 May 2008.
2. Hildebrand and Urquhart (2004).
3. Wilson (2004), pp. 7–8.
4. Wilson (2004), pp. 7–8.
5. Wilson (2004), p. 8.

Bruderheim 8

1. The Queen Elizabeth II Planetarium opened on 22 September 1960, several years before larger facilities were built in other Canadian cities.
2. Sherritt Gordon's nickel refinery began processing ore from the company's mine at Lynn Lake, Manitoba, in 1954. In 1966, a mint was opened, to produce coin blanks and medallions; the first medallion produced commemorated Anthony Henday (who figures in our story, see Chapter 2), the first white man to visit Alberta.
3. Folinsbee and Bayrock (1961), p. 220.
4. Folinsbee and Bayrock (1961), pp. 220–221.
5. Miller (1960), p. 11.
6. Author's interview with Stan Walker, Calgary, Alberta, 26 June 2004.
7. Interview with Stan Walker.
8. By 2005, the University of Alberta had reacquired this specimen for its meteorite collection and loaned a 1.5 kg specimen to the TELUS World of Science Edmonton (the current name for the Queen Elizabeth II Planetarium's successor) for display.
9. Unfortunately, the high school suffered a fire in 1986, and this specimen was not found when the charred rubble was cleared up.
10. Interview with Stan Walker.
11. This comes from a public statement or press release [likely R.E. Folinsbee's] that was undated and unsigned but obviously written shortly after the fall of the meteorite. This and subsequent extracts from Folinsbee's correspondence are used with permission from the Department of Earth and Atmospheric Sciences, University of Alberta.
12. Milton (1960).

13. This comes from a public lecture on meteorites R.E. Folinsbee presented at the Provincial Museum of Alberta on 22 March 1984, on the occasion of the repatriation of the Iron Creek meteorite.
14. Baadsgaard et al. (1961), p. 3577.
15. E. Milton (1960).
16. Extract from a letter to R.E. Folinsbee from S.C. Robinson (11 March 1960).
17. Krinov (1960a).
18. Extract from a letter to S.C. Robinson from R.E. Folinsbee (31 March 1960).
19. Extract from a letter to S.C. Robinson.
20. Folinsbee and Bayrock (1961), p. 224.
21. Folinsbee and Bayrock (1961), p. 224.
22. This comes from a talk on the Bruderheim meteorite R.E. Folinsbee gave to the Rotary Club, 18 January 1961.
23. Extract from a letter to N. Broda from R.E. Folinsbee (6 May 1960).
24. Extract from a letter to R.E. Folinsbee from F. Alexandruk (21 March 1960).
25. Extract from a letter to R.E. Folinsbee from J. Grant (7 September 1960).
26. The final tally was 303 kg, the largest total mass ever for a Canadian meteorite fall or find.
27. Tatum (1990).
28. Folinsbee and Bayrock (1961), p. 226.
29. Salvatori et al. (1984).
30. Extract from a letter to Ward's Natural Science Establishment, Inc. from R.E. Folinsbee (18 March 1960).
31. Extract from a letter to R.E. Folinsbee from Ward's Natural Science Establishment, Inc. (22 March 1960).
32. Extract from a letter to D.A. Taylor from Ward's Natural Science Establishment, Inc. (26 July 1960).
33. This comes from a memorandum, or policy statement, entitled "The Bruderheim Meteorite" on meteorite exchanges, signed by R.E. Folinsbee. The University of Alberta Meteorite Collection is now second only to the National Meteorite Collection in Ottawa.
34. Extract from a letter from Colleen Dougherty (24 April 1963).
35. Extract from a letter from R.E. Folinsbee to Colleen Dougherty (2 May 1963).
36. Edmonton Journal (1960).
37. Baadsgaard et al. (1961), p. 3575.
38. Extract from a form letter from R.E. Folinsbee to "all investigators" (June 1961). This letter was sent to all recipients of Bruderheim samples from the initial distribution. In the event, several papers were published together in the October 1961 *Journal of Geophysical Research*, Vol. 66, No. 10.
39. Baadsgaard et al. (1961).
40. Extract from a letter to M. Calvin from R.E. Folinsbee (7 June 1961).
41. Campbell and Baadsgaard (1961).
42. Mason (1963), p. 1014.
43. Van Schmus and Ribbe (1968).
44. Duke et al. (1961).
45. Smith and Launspach (1991).
46. Smith and Launspach (1991), p. 89.
47. Olsen and Bunch (1984).
48. Dodd and Jarosewich (1979).

49. Stöffler et al. (1991).

50. Egan et al. (1973).

51. Gradie et al. (1980).

52. Normal reflectance, or *albedo* (from the Latin *albus* for *white*), is a measure of a surface's relative brightness when illuminated and observed vertically. Albedo is expressed as the fraction of light that is reflected, ranging from 0, for a perfectly black surface, to 1, for a totally reflective surface; charcoal has an albedo of about 0.04, whereas fresh snow is close to 1.

53. Vogt and Ehmann (1965).

54. Baadsgaard et al. ((1961).

55. Duke et al. (1961).

56. Schmitt et al. (1972).

57. Shima and Honda (1967b).

58. Masuda et al. (1973).

59. Setser and Ehmann (1964), pp. 769–782.

60. Setser and Ehmann (1964), p. 780. Ehmann, in a later paper, subsequently retracted the Zr data (see Ganapathy et al. 1976, p. 305).

61. Ganapathy et al. (1976), p. 302.

62. Ganapathy et al. (1976).

63. Sen Gupta (1968).

64. Allen and Mason (1973). See particularly Figures 1, 2, and 4.

65. Suess and Urey (1956).

66. Ehmann et al. (1961).

67. Ehmann (1965).

68. Tektites are small, greenish, brown, or black glassy blobs thought to be the "splash" or ejecta from an asteroid or comet impact on the Earth that is blasted into space and then re-enters the atmosphere to strike the surface.

69. Amiruddin and Ehmann (1962).

70. Rushbrook and Ehmann (1962).

71. Baedecker and Ehmann (1965).

72. Müller et al. (1971), p. 1130.

73. Imamura and Honda (1976).

74. Tandon and Wasson (1968).

75. Sears and Weeks (1986).

76. Greenland and Goles (1965).

77. Nishimura and Sandell (1964).

78. In July 1960, Donald A. Taylor, the university's museum curator, negotiated an exchange of 3859 g of Bruderheim specimens for $115.50 worth of mineralogical specimens from Ward's catalogue. Unfortunately, three months later Ward's accounting department was still demanding payment. Professor Folinsbee had to write a letter: "It is quite impossible for us to consider providing Ward's with future specimens of Bruderheim until this matter is cleared up."

79. Goles and Anders (1962).

80. The uranium data in Table 2 on p. 730 of Goles and Anders (1962) are a factor of 10 too high; see erratum in *Geochim. Cosmochim. Acta* (1963), p. 981 for details.

81. Smith et al. (1977).

82. The Beckman DU spectrophotometer has proven to be one of the most versatile and widely used analytical instruments in history. From its introduction in 1941 to the end

of production in 1975, over 35,000 instruments using the optical design of the original DU were sold.

83. Shima (1964).
84. Weber et al. (1983).
85. Schultz and Kruse (1989).
86. Sill and Willis (1962).
87. Nagai et al. (1987).
88. Zhai and Shaw (1994).
89. Zhai et al. (1996).
90. Kothari and Goel (1974).
91. Reed and Allen (1966).
92. Wyttenbach et al. (1965).
93. von Gunten et al. (1965).
94. Reed (1964).
95. Allen and Clark (1977).
96. Masuda et al. (1972).
97. Cumming (1974).
98. Crozaz (1979).
99. Shima and Thode (1961).
100. Krouse and Folinsbee (1964).
101. Taylor et al. (1965).
102. The symbol for per cent (%, parts per hundred) is well known to readers, but the less familiar symbol "‰" means parts per thousand.
103. Suess and Wänke (1962).
104. Suess and Wänke (1962), p. 479.
105. Brown et al. (1984).
106. Turner et al. (1966).
107. Matsui et al. (1986).
108. Shima and Honda (1967a).
109. Although ^{87}Rb is, technically speaking, radioactive, it has such a long half-life that it is often listed as a stable isotope. Rubidium actually has five other unstable isotopes, with half-lives ranging from minutes to days, that are of no concern in Rb-Sr dating.
110. Gale et al. (1980).
111. Clarke and Thode (1961).
112. Manuel and Rowe (1964).
113. Merrihue (1963).
114. Merrihue (1966).
115. Merrihue (1966), p. 269.
116. Alexander and Manuel (1969).
117. Clarke and Thode (1964).
118. Merrihue (1966).
119. Ganapathy and Anders (1973).
120. Hodgson and Baker (1969), p. 945.
121. Yang and Epstein (1983), p. 2199.
122. The Kelvin temperature scale (K) was developed by the physicist Lord Kelvin in the mid-1800s. The Kelvin (K; Kelvins, *not* degrees) temperature scale is similar to the Celsius scale, except that it starts at absolute zero (0 K; –273 °C) rather than the freezing point of water (0 °C; 273 K); Kelvins and Celsius degrees are the same size. The Kelvin scale is used mainly by scientists.

123. Fireman and DeFelice (1961).
124. Fireman and DeFelice (1961), p. 3550.
125. Bainbridge et al. (1962).
126. Charalambus and Goebel (1962).
127. The Geiger-Müeller counter is a very sensitive detector of ionizing radiation, but all it does is count any ionizing radiation that passes through it whether it be an alpha particle, proton, electron, or photon. A proportional counter, on the other hand, can distinguish between alpha particles, beta particles, and photons and can be adjusted to count only the alpha particles having more than a certain energy.
128. Rowe and Van Dilla (1961), pp. 3553–3554.
129. Rowe and Van Dilla (1961), pp. 3555–3556.
130. Rowe et al. (1963), p. 988.
131. Honda et al. (1961), pp. 3541–3542.
132. Honda et al. (1961), pp. 3543–3546.
133. Müller et al. (1981).
134. Englert and Herr (1978).
135. Hidaka et al. (2000).
136. Cameron and Top (1975).
137. Lavrukhina and Ustinova (1972).
138. Cameron and Top (1974).
139. Ustinova et al. (1988).
140. Shedlovsky et al. (1967), p. 5053.
141. Cressy (1970), p. 776.
142. Cressy (1971).
143. Bogard and Cressy (1973).
144. Jull et al. (2000).
145. Lavielle et al. (1997).

Ferintosh 9

1. The location of the find is latitude 52°48′N, longitude 112°59′W. In the system of land descriptions on the Canadian prairies, the site is identified as Section 16, Township 44, Range 21, west of the Fourth Meridian.
2. Folinsbee (1966). Sam Enarson's son Dale has the same initials as his cousin Don and was also a University of Alberta student at the time, so it is not clear from Folinsbee's report who brought in the meteorite. However, Dale Enarson, who still farms on the same land, kindly cleared up the confusion for the author.
3. D.A. Enarson, interview by R.E. Folinsbee, Edmonton, 10 November 1965. This and subsequent extracts from Folinsbee's correspondence are used with permission from the Department of Earth and Atmospheric Sciences, University of Alberta.
4. This comes from R.E. Folinsbee's note on his trip, with C.R. Stelck, to Ferintosh on 11 November 1965.
5. Smith (1997), p. 121.
6. Extract from a letter to R.E. Folinsbee from E.L. Krinov (1 December 1965).
7. Smith (1997), pp. 122–123.
8. Smith (1997), p. 125.

9. Smith (1997), pp. 123–125.
10. Smith (1997), p. 125.

Innisfree

1. Halliday et al. (1978), pp. 26–27.
2. Halliday et al. (1978), p. 27.
3. Halliday et al. (1978), pp. 27–28.
4. Halliday et al. (1978), pp. 31–32.
5. Halliday et al. (1978), pp. 34.
6. Halliday et al. (1981), pp. 155–156.
7. Although now in Pisces, the so-called *first point of Aries* marks the position of the Sun in the sky at the vernal (spring) equinox.
8. Halliday et al. (1981), p. 153.
9. Halliday (1987). An unsuccessful search was made in 1987 near Ridgedale by a team from the University of Saskatchewan.
10. Graham (1978).
11. Smith (1980).
12. Smith (1980), p. 440.
13. Kallemeyn et al. (1989).
14. Graham et al. (1985).
15. Rubin (1990).
16. Rubin (1990), p.1231.
17. Rubin (1994).
18. Rubin (1994), p. 93.
19. Rubin (2003).
20. Zhai and Shaw (1994).
21. Zhai et al. (1996).
22. Rancitelli and Laul (1977).
23. Lavrukhina and Gorin (1979).
24. Heusser et al. (1978).
25. O. Müller et al. (1981).
26. Goswami et al. (1978).
27. Melcher and Sears (1979).
28. Melcher and Sears (1979), p. 252.

Peace River

11

1. Folinsbee and Bayrock (1964), p. 110.
2. Folinsbee and Bayrock (1964), pp. 109–110.
3. Folinsbee and Bayrock (1964), pp. 111–113.
4. Folinsbee and Bayrock (1964), p. 118.
5. Folinsbee and Bayrock (1964), p. 120.
6. Folinsbee and Bayrock (1964), pp. 118–119.

7. Folinsbee and Bayrock (1964), p. 119.
8. Baadsgaard et al. (1964), p. 4200.
9. Folinsbee and Bayrock (1964), p. 122.
10. Baadsgaard et al. (1964), p. 4197.
11. Baadsgaard et al. (1964), pp. 4197–4198.
12. A polymorph of a mineral has the same chemical composition as the mineral but a different atomic structure. The atomic structure has changed in response to a change in conditions (e.g., high pressure) and makes the mineral more stable. Another example is *stishovite,* which has an atomic structure much denser than the atomic structure of quartz and, hence, is more stable than quartz in high-pressure conditions (see Chapter 17).
13. Price et al. (1982), p. 729.
14. Price et al. (1983).
15. Price (1983).
16. Chen and El Goresy (2000).
17. Chen and El Goresy (2000), p. 495.
18. Van Schmus and Ribbe (1968), p. 1330.
19. Dodd and Jarosewich (1979).
20. Schmitt et al. (1972), p. 149.
21. Vogt and Ehmann (1965).
22. Precision and accuracy are often confused but do *not* mean the same thing. Precision is a measurement of how closely the analytical results can be duplicated. Accuracy measures how close to a true or accepted value a measurement lies. The usual way to illustrate the difference is to image a dart board: A *precise* dart-thrower will place his darts consistently within a small area, but not necessarily the bull's eye; an *accurate* thrower may scatter his darts all over, but some will hit the bull's eye. Ideally, of course, one hopes for both precision and accuracy in scientific analysis as well as in darts.
23. Taylor (1964b).
24. Shedlovsky et al. (1967).
25. Shedlovsky et al. (1967), p. 5052.
26. Cressy (1970).
27. Sen Gupta (1968).
28. Müller et al. (1971).
29. Greenland and Goles (1965).
30. Shima and Honda (1967b).
31. Masuda et al. (1973), p. 243.
32. Imamura and Honda (1976).
33. Tandon and Wasson (1968).
34. Chou and Cohen (1973).
35. Chou and Cohen (1973), p. 317.
36. Shima and Honda (1967a).
37. Shima and Honda (1967a), p. 341.
38. Cumming (1974).
39. Cumming (1974), p. 259.
40. Schultz and Kruse (1989).
41. These carbon atoms are of the familiar ^{14}C radioisotope variety, elsewhere employed in radioactive carbon dating of archaeological specimens.
42. Kothari and Goel (1974).
43. Allen and Clark (1977).

44. Zhai and Shaw (1994).
45. Zhai et al. (1996).
46. So-called because, until Friedrich Wöhler synthesized urea in 1828 from purely inorganic materials, scientists thought carbon-containing compounds were produced only by life, i.e., *organisms*.
47. Hodgson and Baker (1969).
48. Krouse and Folinsbee (1964).
49. Rowe et al. (1965).
50. Merrihue (1963)
51. Englert and Herr (1978).
52. McConville et al. (1985), p. 707.
53. McConville et al. (1988).
54. Turner et al. (1990), p. 416.
55. Nagai et al. (1987).
56. Nagai et al. (1993).
57. Taylor (1964a).

Skiff   12

1. Steiner and Smith (1980), p. 239.
2. The location of the find is latitude 49°27′N, longitude 111°52′W.
3. The transaction was somewhat convoluted; Folinsbee's secretary prepared an invoice, ostensibly *from* Bill Nemeth *to* the Department of Geology, itemized as "Meteorite (SKIFF)... $300.00," and sent this to Mr. Nemeth who mailed it back for payment.
4. Steiner (1981).

Vilna   13

1. Folinsbee et al. (1969).
2. Folinsbeeet al. (1969), pp. 61–62.
3. Bayrock (1967), p. 5.
4. Bayrock (1967), p.10.
5. Folinsbee et al. (1969), p. 62.
6. Folinsbee et al. (1969), p. 64.
7. Extract from a letter to L.A. Bayrock from Mrs. C.L. Brickman (11 February 1967).
8. Folinsbee et al. (1969), p. 66.
9. Folinsbee et al. (1969), p. 66.
10. Folinsbee et al. (1969), p. 67.
11. Folinsbee et al. (1969), p. 78.
12. Folinsbee et al. (1969), p. 73.
13. Folinsbee et al. (1969), p. 69.
14. Folinsbee et al. (1969), p. 79.
15. Folinsbee (1967).
16. Folinsbee et al. (1969), pp. 79–80.
17. Smith et al. (1973).

1. The location of the fall is latitude 50°31′N, longitude 113°8′W. In the system of land descriptions on the Canadian prairies, the site is identified as Section 17, Township 18, Range 23, west of the Fourth Meridian.
2. The University of Alberta was established in 1906 in Edmonton, the provincial capital; however, a campus was also created in Calgary in 1945. This satellite campus, known as the University of Alberta at Calgary, operated until 1966, when it became the autonomous University of Calgary.
3. This comes from T.A. Oliver's report on the Vulcan meteorite that R.E. Folinsbee forwarded to E.L. Krinov, editor of the *Meteoritical Bulletin,* in Moscow and from Folinsbee's accompanying letter to Krinov (28 September 1964). According to M.H. Hey (1966), the Vulcan report was published in the *Meteoretical Bulletin*, No. 32 (Oliver 1964), but I have been unable to find a copy.
4. Mason (1967).
5. Cameron and Top (1974), p. 899.
6. Cameron and Top (1975).
7. Cameron and Top (1974), p. 904.
8. The author actually found several pages of calculations for the proportions of ingredients needed for a "Vulcan mock-up" among Folinsbee's papers, not a neatly typed up recipe card.

1. Goswami et al. (1980), p. 296.
2. Griffin et al. (1992).
3. Edmonton Journal (1952).
4. Griffin et al. (1992), pp. 5–6.
5. Griffin et al. (1992), p. 6.
6. Griffin et al. (1992), p. 8.
7. Griffin comments: "The name Abee is frequently mispronounced: the correct usage in the local area rhymes with 'baby.'"
8. Griffin et al. (1992), p. 9.
9. R. Kempton (1996b).
10. Anders (1964, p. 634), divided enstatite chondrites into two subclasses: "Type I comprises the three meteorites with the highest iron and sulfur content: Indarch, Abee, and St. Sauveur; type II includes the remaining enstatite chondrites..." This Type I/II scheme was not widely adopted and has been superceded by the EH/EL designations (see Sears et al. 1983; Rubin and Scott 1997).
11. Dawson et al. (1960).
12. Dawson et al. (1960), p. 127.
13. *Breccia* means *broken stones* or *rubble* in Italian and *polymict* means that the fragments came from different sources having different compositions; *monomict* means that the fragments all came from the same source and have the same composition.
14. Urey and Mayeda (1959).
15. Marti (1983), pp. 116–117.

16. Rubin and Keil (1983).

17. Rubin, and Keil (1983), p. 121. They note: "Because it is probable that Dawson et al. originally used water when cutting the Abee slab, much of the oldhamite near the slab's surface may have been dissolved."

18. Sugiura and Strangway (1983).

19. Sugiura and Strangway (1981).

20. Sears (1983).

21. To a chemist, oxidation and reduction mean the exchange of electrons between elements. Oxidation is defined as the loss of electrons; reduction is defined as the gain of electrons. For example, an oxygen atom (O) has an intense affinity for electrons and will "steal" two from another element, becoming reduced and doubly negatively charged (O^{2-}), whereas the atom of the other element (e.g, Fe) is oxidized, losing two electrons and becoming doubly positively charged (Fe^{2+}). An element in its natural state is defined as having an electrical charge equal to zero; hence, the metallic iron present in Abee is denoted Fe^0.

22. Rubin and Scott (1997).

23. Rubin et al. (1997).

24. Rubin and Scott (1997), p. 429.

25. Herndon and Rudee (1978).

26. Kempton (1996a).

27. Kempton (1996b).

28. Kempton (1996a), p. 19.

29. R. Herd, email message to author, 20 October 2006.

30. Keil (1968).

31. Zhang et al. (1996).

32. Frazier and Boynton (1985), p. 198.

33. Leitch and Smith (1982).

34. Leitch and Smith (1982), p. 2086. They also describe Abee as a *monomict* breccia, in apparent contradiction of the observations of others (e.g., Dawson et al. 1960; Searset al. 1983).

35. Shimizu et al. (2002).

36. Douglas and Plant (1968).

37. Olsen et al. (1973).

38. Dollfus et al. (1980).

39. Binzel (2001).

40. von Michaelis et al. (1969a).

41. von Michaelis et al. (1969b).

42. See Chapter 11, note 22, for a discussion of precision and accuracy.

43. Schmitt et al. (1972).

44. Mason (1971).

45. Ikramuddin et al. (1976).

46. Ikramuddin et al. (1979).

47. Biswas et al. (1980).

48. Laul et al. (1973).

49. See Chapter 8, note 122, for a description of the Kelvin temperature scale.

50. Alaerts and Anders (1979).

51. Schmitt et al. (1963a).

52. Evensen et al. (1978).

53. McCulloch et al. (1977).

54. Nakamura and Masuda (1973).

55. Frazier and Boynton (1980).

56. Frazier and Boynton (1985), p. 198.

57. Loveless (1970).

58. Loveless et al. (1972).

59. Greenland and Lovering (1965).

60. Greenland and Lovering (1965), p. 847.

61. Shima and Honda (1967b).

62. Sill and Willis (1962).

63. Binz et al. (1974).

64. Shaw (1974).

65. Fouche and Smales (1966).

66. Anders (1964). See note 10 above.

67. Rambaldi and Cendales (1980).

68. Hertogen et al. (1983).

69. One AU, or *astronomical unit*, is the average distance between the Earth and the Sun ($\sim 1.5 \times 10^8$ km).

70. Hertogen et al. (1983).

71. Setser and Ehmann (1964).

72. Greenland and Goles (1965).

73. Nishimura and Sandell (1964).

74. Schmitt et al. (1963b).

75. Baedecker and Ehmann (1965).

76. Baedecker, and Wasson (1965).

77. Sen Gupta (1968).

78. Shima (1964).

79. Vogt and Ehmann (1965).

80. Curtis et al. (1980).

81. Zhai and Shaw (1994).

82. Zhai et al. (1996).

83. Murrell and Burnett (1982).

84. Masuda et al. (1972).

85. Woolum et al. (1979).

86. Smith et al. (1977).

87. Smith and De Laeter (1986).

88. Russell et al. (1978).

89. Niederer et al. (1985).

90. Goles and Anders (1962).

91. Schultz and Kruse (1989).

92. Merrihue (1966).

93. Clarke and Thode (1964).

94. Jeffery and Reynolds (1961).

95. Srinivasan et al. (1978).

96. Crabb and Anders (1982).

97. Reed et al. (1960).

98. Hintenberger et al. (1973).

99. Imamura and Honda (1976).

100. Reed (1964).
101. Reed and Allen (1966).
102. von Gunten et al. (1965).
103. Thiemens and Clayton (1983).
104. Thiemens and Clayton (1983), p. 167.
105. By convention, isotopic compositions are represented by the Greek letter, δ, and are calculated relative to a standard, for example,

$$\delta^{13}C = \frac{(^{13}C/^{12}C)^{sample} - (^{13}C/^{12}C)^{standard}}{(^{13}C/^{12}C)^{standard}} \times 1000$$

106. Grady et al. (1986).
107. Belsky and Kaplan (1970).
108. Yang and Epstein (1983), p. 2199.
109. Miller (1953).
110. Although Urey was Miller's research advisor and the two worked together on the experiment, when Miller showed a draft of his paper to Urey, he was surprised by the latter's immediate and generous response. Urey said his name should not be on the paper because if it were, Miller would receive little or no credit. The Nobel Laureate Urey also called the editor of *Science* and asked that the paper be published as soon as possible.
111. Russell et al. (1992).
112. Ott et al. (1992).
113. By comparison, the ore mined at the Diavik diamond mine in the Northwest Territories contains an average of 4 carats per tonne, or 0.8 ppm by mass.
114. Deines and Wickman (1985).
115. Deines and Wickman (1985), pp. 93–94.
116. Taylor et al. (1965).
117. Reuter et al. (1965).
118. Shima and Thode (1961).
119. Kaplan and Hulston (1966).
120. Bogard et al. (1983).
121. Wacker and Marti (1983).
122. Goswami (1983).
123. Goswami (1983), p. 163.
124. Rowe et al. (1963), p. 988.
125. Englert and Herr (1978).
126. Cameron and Top (1975).
127. Cameron and Top (1974).
128. Cressy (1971).
129. Shima and Honda (1967a).

Great Balls of Fire 16

1. The geographic location of the Meanook Geomagnetic Observatory is latitude 54°36′57″N, longitude 113°21′45″W. The Newbrook Observatory's location is latitude 54°19′27″N, longitude 112°57′15″W.

2. The second International Polar Year, which followed the first (1882–1883) by 50 years, was a co-ordinated international program of scientific research in the Arctic. Although its scope was reduced by the economic exigencies of the Great Depression and its final reporting delayed by World War II, it provided the inspiration for the much more ambitious International Geophysical Year (IGY) of 1957–1958.
3. Millman (1959).
4. Millman (1959), p. 21.
5. Millman (1959), p. 32.
6. Halliday et al. (1978).
7. Halliday et al. (1978), p. 23.
8. Halliday et al. (1989a).
9. Halliday et al. (1989a), p. 72.
10. Halliday et al. (1989b).
11. Halliday et al. (1996).

Stalking the Wild Meteorite   17

1. It seems probable that the Plains Indians had to mount a search for the Manitou Stone, or Iron Creek meteorite, which fell so spectacularly in pre-contact times; it is unlikely to have landed within sight of the then sparse population.
2. Stardust Magazine (1955a), p. 1. Stardust Magazine is the newsletter of the Edmonton Centre of the Royal Astronomical Society of Canada.
3. Stardust Magazine (1955b), p. 1.
4. Halliday (1985).
5. Folinsbee et al. (1986).
6. Brown and Zalcik (1992). This investigation of MORP sites was sponsored by the Associate Committee on Meteorites (ACOM).
7. LaPaz (1946).
8. Bill 40, 2nd Session, 14th Legislature, Alberta 9 Elizabeth II, 1961. Government of Alberta, Edmonton.
9. *Cultural Property Export and Import Act,* Chapter C-51 and Canadian Cultural Property Export Control List. Government of Canada, Ottawa.
10. Schmitt (2002).
11. Millman (1964).

The Dinosaur Killer   18

1. Alvarez et al. (1980).
2. Hildebrand (1993).
3. The crater was named Chicxulub (variously pronounced "Chic-sa-lube" or "Chic-su-lub") by Hildebrand after a village near the centre of the crater.
4. The Shoemaker-Levy 9 comet was so-named because it was the ninth comet discovered by the team of Eugene and Carolyn Shoemaker and David Levy. The late Eugene Shoemaker (1928–1997) pioneered the study of impact craters as a scientific

discipline while working at the United States Geological Survey and with NASA. Carolyn Shoemaker (Palomar Observatory) is the most successful comet discoverer in history with over 30 to her credit, as well as more than 800 asteroids. David Levy, an amateur astronomer and long-time member of the Royal Astronomical Society of Canada, was born in Montreal and now lives in Arizona. He has discovered more than 20 comets.

5. Lerbekmo and St. Louis (1986).
6. Carlisle (1995), pp. 17–22.
7. Grieve and Alexopoulis (1988).
8. For comparison, the Manicouagan crater in Quebec, which appears as a conspicuous ring-shaped lake (and can be seen on television weather maps of Quebec), is 214 km in diameter.
9. The *Observer's Handbook*, published annually by the Royal Astronomical Society of Canada (Toronto), provides an authoritative listing of all confirmed impact craters in North America. The University of New Brunswick maintains an online database of confirmed impact structures from around the world, at http://www.unb.ca/passc/ImpactDatabase/
10. A *simple* crater is a simple bowl-shaped depression with a raised rim, 2–4 km wide, formed by lower-energy events. A *complex* crater is the result of a much larger impact; it may be up to several tens of kilometres in diameter and usually has a central uplift region and a large surrounding ejecta blanket. At still larger impact energies, the central lift region is no longer a single mound or peak but instead forms an inner ring inside the outer rim; Chicxulub is an example of a multi-ring terrestrial crater.
11. Haites and van Hees (1962).
12. Sawatzky (1976).
13. Hanova et al. (2005).
14. Lawton, D.C., R.R. Stewart, and R. Gault. (1993). The Geophysical Expression of the Eagle Butte Impact Structure. Paper presented at the 19th Annual Meeting of the Canadian Geophysical Union, Banff, Alberta. Shatter cones are conical fragments of rock with striations flaring out from the apex towards the base of the cone; they are produced only by the intense shock pressures created by impact events and nuclear explosions.
15. Mazur and Stewart (1998).
16. Mazur (1999), p. 74 , fig. 4.1.1. This is the geographical location of the oil or gas well identified as "15–27–98–06W6" and considered by Mazur as marking the centre of the Hotchkiss structure.
17. Mazur and Stewart (1998), p. 35.3.
18. Isaac and Stewart (1993).
19. Westbroek (1997).

Epilogue 19

1. Halliday (2001).
2. A. Zalaski, note to author, 2006.
3. Alan Zalaski identifies these sections as Township 56, Range 8, Sections 5 and 6; Township 55, Range 8, Sections 24 and 29.
4. Extract from a letter to M. Luski from C. Herd (5 September 2007).

5. Britt and Consolmagno (2003).
6. Garrison (2000).
7. Hsu (1998).
8. Love and Keil (1995).
9. Okazaki et al. (2000).
10. Patzer and Schultz (2001).
11. Patzer and Schultz (2002).
12. Brazzle et al. (1999).
13. Ceplecha and Revelle (2005).
14. Rubin (2004).
15. Reisener and Goldstein (2003).

Glossary

alabandite (MnS): a sulphide mineral containing manganese.

α-cristobalite (SiO_2): a polymorph of quartz.

amphibole: a general term for the group of silicate minerals characterized by a structure consisting of double chains of silica tetrahedra and that typically contain water.

anomaly: any difference in the abundance of an element, isotope, etc. relative to its abundance in some reference material, e.g., terrestrial rocks.

apatite [$Ca_5(PO_4)_3(OH,F,Cl)$]: a phosphate mineral containing calcium plus varying proportions of hydroxyl groups, fluorine, and chlorine.

breccia: angular fragments, or pieces, of rock; a brecciated meteorite is made up of cemented breccia.

bronzite: a name for calcium-poor pyroxene; also a redundant or obsolete name for a chondrite containing this mineral.

cohenite [$(Fe,Ni,Co)_3C$]: a carbide mineral commonly found in iron meteorites.

daubreelite ($FeCr_2S_4$): a sulphide mineral containing chromium.

diaplectic: a glass-like substance formed by shock metamorphism; it has the same composition as the original mineral but a lower density.

djerfisherite ($K_6Na(Fe,Cu,Ni)_{25}S_{26}Cl$): a rare sulphide mineral.

enstatite ($MgSiO_3$): a variety of *pyroxene* with relatively little or no iron.

equilibrated chondrite: a chondrite of petrologic type 4 to 7; in such a chondrite the major silicate mineral has a uniform composition.

excentroradial: a type of chondrule in which the pyroxene crystals appear to radiate from a single, off-centre point.

facies: a set of characteristics, such as grain size, density, etc., that distinguish a given section of rock from nearby sections.

fall: a meteorite that is recovered shortly after its fall is witnessed.

fayalite (Fe_2SiO_4): the iron-rich end-member of the olivine *solid solution series;* abbreviated *Fa.*

ferrosilite ($FeSiO_3$): a variety of *pyroxene* with relatively little or no magnesium.

find: a meteorite that is found, often by chance, long after it has fallen unseen.

forsterite (Mg_2SiO_4): the magnesium-rich end-member of the *olivine* solid solution series; abbreviated *Fo.*

giga annum (Ga): a unit of time equal to 10^9 years (one billion on the short scale, one milliard on the long scale).

goethite [$Fe^{3+}O(OH)$]: a mineral form of iron oxide, i.e., "rust."

hypersthene: the intermediate member in the enstatite-ferrosilite solid solution series with around 50% iron; also a redundant or obsolete name for a chondrite containing this mineral.

ilmenite [$(Fe,Mg)TiO_3$]: this mineral is the most important terrestrial ore of titanium and also has been found in meteorites.

keilite [$(Fe,Mg)S$]: a sulphide mineral first identified in the Abee meteorite.

lithophile: an element that tends to concentrate in the silicate rather than in the metal or sulphide phases of meteorites.

maskelynite: a type of natural glass created by the vitrification of *plagioclase* by shock melting from meteorite impact.

mega annum (Ma): a unit of time equal to one million (10^6) years.

merrillite [$Ca_9MgH(PO_4)_7$]: a calcium phosphate mineral, synonymous with whitlockite.

modal analysis: the manual or automated identification and counting of mineral grains in a systematic, grid-wise examination of a meteorite thin section using a range of techniques including optical microscopy and electron microprobe analysis.

niningerite [$(Mg,Fe^{2+},Mn)S$]: a sulphide mineral named after meteoriticist Harvey H. Nininger.

oldhamite [$(Ca,Mg,Fe)S$]: a calcium-rich sulphide mineral.

olivine: not a single mineral, but a solid solution series between magnesium-rich forsterite and iron-rich fayalite.

phase: a collective noun used by meteoriticists to denote all those portions of, for example, a meteorite sample, that are identical in chemical composition and physical state, e.g., "the troilite phase."

phenocryst: a large, well-formed crystal embedded in a fine-grained or glassy groundmass.

polymorph: a mineral with a single chemical composition can exist with two or more different crystal structures, or polymorphs, depending upon the pressure and temperature the mineral is exposed to.

pyroxene: not a single mineral, but a group of several minerals characterized by a structure consisting of single chains of silica tetrahedra.

rare earth elements (REE): elements 57 (lanthanum) to 103 (lawrencium) in the periodic table. They are composed of two series: the lanthanides (lanthanum to lutetium) and the actinides (actinium to lawrencium).

richterite [$Na(CaNa)(Mg,Fe^{2+})5[Si_8O_{22}](OH,F)_2$]: a sodium-rich silicate mineral of the amphibole group.

ringwoodite: a polymorph of olivine.

schreibersite [$(Fe,Ni)_3P$]: a phosphate mineral.

shock metamorphism: the irreversible chemical and physical changes produced in a rock or mineral by the extreme pressures and temperatures created in an impact event.

siderophile: an element that normally associates with iron.

sinoite (Si_2N_2O): a rare mineral associated with enstatite in chondrites.

solid solution series: occurs when two or more elements can substitute for each other in a crystal structure without much alteration of the structure.

wadsleyite ($(Mg,Fe)_2SiO_4$): a polymorph of olivine discovered in the Peace River meteorite.

whitlockite: a calcium phosphate mineral, synonymous with merrillite.

References

Alaerts, L., and E. Anders. 1979. On the kinetics of volatile loss from chondrites. Geochim. Cosmochim. Acta, 43: 547–553

Alexander, E.C., and O.K. Manuel. 1969. Xenon in chondrules and matrix material of chondrites. Geochim. Cosmochim. Acta, 33: 298–301.

Allan, J.A. Edmonton meteorite. Report, Edmonton, 1953.

Allen, R.O., and P.J. Clark. 1977. Fluorine in meteorites. Geochim. Cosmochim. Acta, 41: 581–585.

Allen, R.O., and B. Mason. 1973. Minor and trace elements in some meteoritic minerals. Geochim. Cosmochim. Acta, 37: 1435–1456.

Alvarez, L., W. Alvarez, F. Asaro, and H.V. Michel. 1980. Extraterrestial cause for the Cretaceous –Tertiary extinction. Science (Washington, DC), 208: 1095–1108.

Amiruddin, A., and W.D. Ehmann. 1962. Tungsten abundances in meteoritic and terrestrial materials. Geochim. Cosmochim. Acta, 26: 1011–1022.

Anders, E. 1964. Origin, age, and composition of meteorites. Space Sci. Rev. 3: 583–714.

Aylmer, D., V. Bonanno, G.F. Herzog, H. Weber, J. Klein, and R. Middleton. 1988. ^{26}Al and ^{10}Be production in iron meteorites. Earth Planet. Sci. Lett. 88: 107–118.

Baadsgaard, H., F.A. Campbell, and R.E. Folinsbee. 1961. The Bruderheim meteorite. J. Geophys. Res. 66: 3574–3577.

Baadsgaard, H., R.E. Folinsbee, and G.L. Cumming. 1964. Peace River meteorite. J. Geophys. Res. 69: 4197–4200.

Baedecker, P.A., and W.D. Ehmann. 1965. The distribution of some noble metals in meteorites and natural materials. Geochim. Cosmochim. Acta, 29: 329–342.

Baedecker, P.A., and J.T. Wasson. 1975. Elemental fractionations among enstatite chondrites. Geochim. Cosmochim. Acta, 39: 735–765.

Bainbridge, A.E., H.E. Suess, and H. Wänke. 1962. The tritium content of three stony meteorites and one iron meteorite. Geochim. Cosmochim. Acta, 26: 471–473.

Bauer, C.A. 1963. The helium contents of metallic meteorites. J. Geophys. Res. 68: 6043–6057.

Bayrock, L.A. 1967. Meteorites—search for witnessed falls in Canada: method, status and aims. Internal Report. to the Associate Committee on Meteorites. National Research Council, Ottawa, ON.

Beech, M., and D. Steel. 1995. On the definition of the term 'meteroid.' Q. J. R. Astron. Soc. 36: 281–284.

Belsky, T., and I.R. Kaplan. 1970. Light hydrocarbon gases, C^{13}, and origin of organic matter in carbonaceous chondrites. Geochim. Cosmochim. Acta, 34: 257–278.

Binz, C.M., R.K. Kurimoto, and M.E. Lipschutz. 1974. Trace elements in primitive meteorites—V. Abundance patterns of thirteen trace elements and interelement relationships in enstatite chondrites. Geochim. Cosmochim. Acta, 38: 1579–1606.

Binzel, R.P. 2001. A new century for asteroids. Sky & Telescope, 102: 44–51.

Biswas, S., T. Walsh, G. Bart, and M.E. Lipshutz. 1980. Thermal metamorphism of primitive meteorites—XI. The enstatite meteorites: origin and evolution of a parent body. Geochim. Cosmochim. Acta, 44: 2097–2110.

Blakeslee, D.J. 2002. Sacred places, sacred symbols: meteoritic iron and the garden of the evening star. *In* Ancient Travellers: Proceedings of the Twenty-Seventh Annual Conference of the Archaeological Association of the University of Calgary. *Edited by* C. Allum, J. Kahn, C. Cluney, and M. Peurakmaki-Brown. The Archaeological Association of the University of Calgary, Calgary, AB. pp. 52–59.

Bogard, D.D., and P.J. Cressy. 1973. Spallation production of ^{3}He, ^{21}Ne, and ^{38}Ar from target elements in the Bruderheim chondrite. Geochim. Cosmochim. Acta, 37: 527–546.

Bogard, D.D., D.M. Unruh, and M. Tatsumoto. 1983. ^{40}Ar/^{39}Ar and U-Th-Pb dating of separated clasts from the Abee E4 chondrite. Earth Planet. Sci. Lett. 62: 132–146.

Brazzle, R.H., O.V. Pravdivtseva, A.P. Meshik, and C.M. Hohenberg. 1999. Verification and interpretation of the I-Xe chronometer. Geochim. Cosmochim. Acta, 63: 739–760.

Britt, D.T., and G.J. Consolmagno. 2003. Stony meteorite porosities and densities: a review of the data through 2001. Meteoritics Planet. Sci. 38: 1161–1180.

Brown, P., and M. Zalcik. 1992. Recent field research on potential meteorite falls from the Meteorite Observation and Recovery Project. J. R. Astron. Soc. Can. 86: 130–139.

Brown, R.M., H.R. Andrews, G.C. Ball, N. Burn. Y. Imahori, J.C.D. Milton, and E.L. Fireman. 1984. ^{14}C content of ten meteorites measured by tandem accelerator mass spectrometry. Earth Planet. Sci. Lett. 67: 1–8.

Buchwald, V.F. 1975. Handbook of iron meteorites. Vols. 1–3. University of California Press, Berkeley, CA.

Butler, W.F. 1872. The great lone land. MacMillan Company of Canada, Toronto, ON.

Cameron, I.R., and Z. Top. 1974. Measurement of Al^{26} in stone meteorites and its use in the derivation of orbital elements Geochim. Cosmochim. Acta, 38: 899–909.

Cameron, I.R., and Z. Top. 1975. A final summary of measurements of cosmogenic Al^{26} in stone meteorites. Geochim. Cosmochim. Acta, 39: 1705–1707.

Campbell, F.A., and H. Baadsgaard. 1961. The Bruderheim meteorite, chemistry and mineralogy. Abstract. J. Geophys. Res. 66 (18): 2518.

Carlisle, D.B. 1995. Dinosaurs, diamonds, and things from outer space: the great extinction. Stanford University Press, Stanford, CA.

Ceplecha, Z., and D.O. Revelle. 2005. Fragmentation model of meteoroid motion, mass loss and radiation in the atmosphere. Meteoritics Planet. Sci. 40: 35–54.

Charalambus, St., and K. Goebel. 1962. Tritium and argon^{39} in the Bruderheim meteorite. Geochim. Cosmochim. Acta, 26: 659–663.

Cheadle, W.B. 1865. Cheadle's journal of trip across Canada 1862–1863. Reprinted in 1931 by Graphic Publishers Limited, Ottawa, ON.

Chen, M., and A. El Goresy. 2000. The nature of maskelynite in shocked meteorites: not diaplectic glass but a glass quenched from shock-induced dense melt at high pressures. Earth Planet. Sci. Lett. 179: 489–502.

Choi, B.-G., X. Ouyang, and J.T. Wasson. 1995. Classification and origin of IAB and IIICD iron meteorites. Geochim. Cosmochim. Acta, 59: 593–612.

Chou, C.L., and A.J. Cohen. 1973. Gallium and germanium in the metal and silicates of L- and LL-chondrites. Geochim. Cosmochim. Acta, 37: 315–327.

Clarke, R.S, and H. Plotkin. 2002. Frederick C. Leonard (1896–1960): first UCLA astronomer and founding father of the Meteoritical Society. Meteoritics Planet. Sci. 37 (Suppl.): A34.

Clarke, W.B., and H.G. Thode. 1961. Xenon in the Bruderheim meteorite. J. Geophys. Res. 66: 3578-3579.

Clarke, W.B., and H.G. Thode. 1964. The isotopic composition of krypton in meteorites. J. Geophys. Res. 69: 3673-3679.

Coleman, A.P. 1886. A meteorite from the northwest. Trans. R. Soc. Can. 4 (3): 97.

Crabb, J., and E. Anders. 1982. On the siting of noble gases in E-chondrites. Geochim. Cosmochim. Acta, 46: 2351-2361.

Cressy, P.J., Jr. 1970. Multiparameter analysis of gamma radiation from the Barwell, St. Séverin and Tatlith meteorites. Geochim. Cosmochim. Acta, 34: 771-779.

Cressey, P.J., Jr. 1971. The production rate of Al^{26} from target elements in the Bruderheim chondrite. Geochim. Cosmochim. Acta, 35: 1283-1296.

Crozaz, G. 1979. Uranium and thorium microdistributions in stony meteorites. Geochim. Cosmochim. Acta, 43: 127-136.

Cumming, G.L. 1974. Determination of uranium and thorium in meteorites by the delayed neutron method. Chem. Geol. 13: 257-267.

Curtis, D., E. Gladney, and E. Jurney. 1980. A revision of the meteorite based cosmic abundance of boron. Geochim. Cosmochim. Acta, 44: 1945-1953.

Dawson, K.R., J.A. Maxwell, and D.E. Parsons. 1960. A description of the meteorite that fell near Abee, Alberta, Canada. Geochim. Cosmochim. Acta, 21: 127-144.

Deines, P., and F.E. Wickman. 1985. The stable carbon isotopes in enstatite chondrites and Cumberland Falls. Geochim. Cosmochim. Acta, 49: 89-95.

Dodd, R.T. 1981. Meteorites: a petrologic-chemical synthesis. Cambridge University Press, Cambridge, UK.

Dodd, R.T., and E. Jarosewich. 1979. Incipient melting in and shock classification of L-group chondrites. Earth Planet. Sci. Lett. 44: 335-340.

Dollfus, A., A. Cilleux, B. Cervelle, C.T. Hua, and J. Mandeville. 1980. Reflectance spectrophometry extended to U.V. for terrestrial, lunar and meteoritic samples. Geochim. Cosmochim. Acta, 44: 1293-1310.

Douglas, J.A.V., and A.G. Plant. 1968. Amphibole: first occurrence in an enstatite chondrite. Abstract. Meteoritics, 4: 166.

Duke, M., D. Maynes, and H. Brown. 1961. The petrography and chemical composition of the Bruderheim meteorite. J Geophys. Res. 66: 3557-3563.

Edmonton Journal. 1952. Meteorite seen falling northeast of Edmonton. Edmonton Journal (10 June). p. 30

Edmonton Journal. 1960. Place meteorite in museum. Edmonton Journal (30 June). p. 12.

Egan, W.G., J. Veverka, M. Noland, and T. Hilgeman. 1973. Photometric and polarimetric properties of the Bruderheim chondritic meteorite. Icarus, 19: 358-371.

Ehmann, W.D. 1965. On some tantalum abundances in meteorites and tektites. Geochim. Cosmochim. Acta, 29: 43-48.

Ehmann, W.D., A. Ambiruddin, P.R. Rushbrook, and M.E. Hurst. 1961. Some trace element abundances in the Bruderheim meteorite. J. Geophys. Res. 66: 3581.

Englert, P., and W. Herr. 1978. A study on exposure ages of chondrites based on spallogenic [53]Mn. Geochim. Cosmochim. Acta, 42: 1635-1643.

Evensen, N.M., P.J. Hamilton, and R.K. O'Nions. 1978. Rare-earth abundances in chondritic meteorites. Geochim. Cosmochim. Acta, 42: 1199-1212.

Farrington, O.C. 1907. Meteorite studies II. Field Museum, Chicago, IL. Field Mus. Nat. Hist. Geol. Ser. 3(6): 113-115; also plate No. 35.

Fireman, E.L., and J. DeFelice. 1961. Tritium, argon 37, and argon 39 in the Bruderheim meteorite. J. Geophys. Res. 66: 3547-3551.

Folinsbee, R.E. 1953. The Edmonton meteorite. Report. University of Alberta, Edmonton.

Folinsbee, R.E. 1964. Discovery of Mayerthorpe iron meteorite, Alberta, Canada. Meteoritical Bull. No. 32.

Folinsbee, R.E. 1965. Meteoritical Bull. No. 33. p. 5.

Folinsbee, R.E. 1966. Discovery of Ferintosh stony meteorite, Canada. Meteoritical Bull. No. 35.

Folinsbee, R.E. 1967. Fall of Vilna stony meteorite, Canada. Meteoritical Bull. No. 40. Meteoritics, 5: 92.

Folinsbee, R.E., and L.A. Bayrock. 1961. The Bruderheim meteorite—fall and recovery. J. R. Astron. Soc. Can. 55: 218–228.

Folinsbee, R.E., and L.A. Bayrock. 1964. The Peace River meteorite: fall and recovery. J. R. Astron. Soc. Can. 58: 109–124.

Folinsbee, R.E., L.A. Bayrock, G.L. Cumming, and D.G.W. Smith. 1969. Vilna meteorite— camera, visual, seismic and analytic records. J. R. Astron. Soc. Can. 63: 61–86.

Folinsbee, R.E., D.G.W. Smith, D.P. Hube, and K J. Finstad. 1986. Two Alberta fireballs in 1985. J. R. Astron. Soc. Can. 80: 312–327.

Fouche, K.F., and A.A. Smales. 1966. The distribution of trace elements in chondritic meteorites. I. Gallium, germanium and indium. Chem. Geol. 2: 5–33.

Frazier, R.M., and W.V. Boynton. 1980. Rare earth element abundances in separates from the enstatite chondrite Abee. Meteoritics, 15: 291.

Frazier, R.M., and W.V. Boynton. 1985. Rare earth and other elements in components of the Abee enstatite chondrite. Meteoritics, 20: 197–218.

Gale, N.H., J.W. Arden, and M.C.B. Abranches. 1980. Uranium-lead age of the Bruderheim L6 chondrite and the 500-Ma shock event in the L-group parent body. Earth Planet. Sci. Lett. 48: 311–324.

Ganapathy, R., and E. Anders. 1973. Noble gases in eleven H-chondrites. Geochim. Cosmochim. Acta, 37: 359–362.

Ganapathy, R., G.M. Papia, and L. Grossman. 1976. The abundances of zirconium and hafnium in the solar system. Earth Planet. Sci. Lett. No. 29: 302–308.

Garrison, D., S. Hamlin, and D. Bogard. 2000. Chlorine abundances in meteorites. Meteoritics Planet. Sci. 35: 419–429.

Goles, G.G., and E. Anders. 1962. Abundances of iodine, tellerium and uranium in meteorites. Geochim. Cosmochim. Acta, 26: 723–737.

Goswami, J.N. 1983. Nuclear track records in the Abee enstatite chondrite. Earth Planet. Sci. Lett. 62: 159–164.

Goswami, J.N., D. Lai, M.N. Rao, N. Sinha, and T.R. Venkatesan. 1978. Particle track and rare gas studies of the Innisfree meteorite. Meteoritics, 13: 481–484.

Goswami, J.N., D. Lai, and N. Sinha. 1980. Nuclear track records in the Abee chondrite. Meteoritics, 15: 295–296.

Gradie, J., J. Veverka, and B. Buratti. 1980. The effects of scattering geometry on the spectrophotometric properties of powdered material. Proc. Lunar Planet. Sci. Conf. 11: 799–815.

Grady, M.M. 2000. The catalogue of meteorites. 5th ed. Cambridge University Press, Cambridge, UK.

Grady, M.M., I.P. Wright, L.P. Carr, and C.T. Pillinger. 1986. Compositional difference in enstatite chondrites based on carbon and nitrogen stable isotope measurements. Geochim. Cosmochim. Acta, 50: 2799–2813.

Graham, A.L. 1978. Fall of the Innisfree, Canada, stony meteorite. Meteoritical Bull. No. 55: 338.

Graham, A.L., A.W. Bevan, and R. Hutchinson. 1985. Catalogue of meteorites. University of Arizona Press, Tucson, AZ.

Greenland, L., and G.G. Goles. 1965. Copper and zinc abundances in chondritic meteorites. Geochim. Cosmochim. Acta, 29: 1285–1292.

Greenland, L., and J.F. Lovering. 1965. Minor and trace element abundances in chondritic meteorites. Geochim. Cosmochim. Acta, 29: 821–858.

Grieve, R.A.F., and Alexopoulis, J. 1988. Microscopic planar features in quartz from Scollard Canyon, Alberta, and the Cretaceous–Tertiary boundary event. Can. J. Earth Sci. 25: 1530–1534.

Griffin, A.A., P.M. Millman, and I. Halliday. 1992. The fall of the Abee meteorite and its probable orbit. J. R. Astron. Soc. Can. 86(1): 5–14.

Haack, H., and E.R.D. Scott. 1993. Chemical fractionations in group IIIAB iron meteorites: origin by dendritic crystallization of an asteroidal core. Geochim. Cosmochim. Acta, 57: 3457–3472.

Haites, T.B., and H. van Hees. 1962. The origin of some anomalies in the plains of western Canada. J. Alta. Soc. Petrol. Geol. 10: 511–533.

Halliday, I. 1985. The Grande Prairie fireball of 1984 February 22. J. R. Astron. Soc. Can. 79: 197–214.

Halliday, I. 1987. Detection of a meteorite "stream": observations of a second meteorite fall from the orbit of the Innisfree chondrite. Icarus, 69: 550–556.

Halliday, I. 2001. The present day flux of meteorites to the Earth. *In* Accretion of extraterrestrial matter throughout Earth's history. *Edited by* B. Peuker-Ehrenbrink and B. Schmitz. Kluwer Academic Publishers/Plenum Press, New York. pp. 305–318.

Halliday, I., A.T. Blackwell, and A.A. Griffin. 1978. The Innisfree meteorite and the Canadian Camera Network. J. R. Astron. Soc. Can. 72: 15–39.

Halliday, I., A.A. Griffin, and A.T. Blackwell. 1981. The Innisfree meteorite fall: a photographic analysis of fragmentation, dynamics and luminosity. Meteoritics, 16(2): 153–170.

Halliday, I., A.T. Blackwell, and A.A. Griffin. 1989a. Detailed records of many unrecovered meteorites in western Canada for which further searches are recommended. J. R. Astron. Soc. Can. 83: 49–80.

Halliday, I., A.T. Blackwell, and A.A. Griffin. 1989b. The typical meteorite event, based on photographic records of fireballs. Meteoritics, 24: 65–72.

Halliday, I., A.A. Griffin, and A.T. Blackwell. 1996. Detailed data for 259 fireballs from the Canadian Camera Network and inferences concerning the influx of large meteoroids. Meteoritics Planet. Sci. 31: 185–217.

Hanova, J., D.C. Lawton, J. Visser, A.R. Hildebrand, and L. Ferriere. 2005. 3D structural interpretation of the Eagle Butte impact structure, Alberta. Universities Space Research Association, Lunar and Planetary Institute, Columbia, MD. Available from http://www.lpi.usra.edu/meetings/lpsc2005/pdf/2355.pdf

Harvey, D.W. 1979. Discovery of the Millarville, Canada, iron meteorite. Meteoritical Bull. No. 56. Meteoritics, 14: 168.

Henry, A., and D. Thompson. 1810. The manuscript journals 1799–1814. Vol. 2. Francis P. Harper, New York.

Herndon, J.M., and M.L. Rudee. 1978. Thermal history of the Abee enstatite chondrite. Earth Planet. Sci. Lett. 41: 101–106.

Hertogen, J., M. Janssens, H. Takahashi, J.W. Morgan and E. Anders. 1983. Enstatite chondrites: trace element clues to their origin. Geochim. Cosmochim. Acta, 47: 2241–2255.

Heusser, G., W. Hampel, T. Kirsten, and O.A. Schaeffer. 1978. Cosmogenic isotopes in recently fallen meteorites. Meteoritics, 13: 492–494.

Hey, M.H. 1966. Catalogue of meteorites. British Museum (Natural History), London.

Hidaka, H., M. Ebihara, and S. Yoneda. 2000. Isotopic study of neutron capture effects on Sm and Gd in chondrites. Earth Planet. Sci. Lett. 180: 29–37.

Hildebrand, A.R. 1993. The Cretaceous/Tertiary boundary impact (Or the dinosaurs didn't have a chance). J. R. Astron. Soc. Can. 87: 77–118.

Hildebrand, A., and D. Urquhart. 2004. Alberta Farmer finds Canada's newest meteorite. University of Calgary, Calgary, AB. In the News (4 June 2004). Available from http://www. ucalgary.ca/oncampus/weekly/june25-04/meteorite.html.

Hintenberger, H., K.P. Jochum, and M. Seufert. 1973. The concentrations of the heavy metals in the four new Antarctic meteorites (a), (b), (c) and (d) and in Orgueil, Murray, Allende, Abee, Allegan, Mocs and Johnstown. Earth Planet. Sci. Lett. 20: 391–394.

Hodgson, G.W., and B.L. Baker. 1969. Porphyrins in meteorites: metal complexes in Orgueil, Murray, Cold Bokkeveld, and Mokoia carbonaceous chondrites. Geochim. Cosmochim. Acta, 33: 943–958.

Honda, M., S. Umemoto, and J.R. Arnold. 1961. Radioactive species produced by cosmic rays in Bruderheim and other stone meteorites. J. Geophys. Res. 66: 3541–3546.

Hsu, W. 1998. Geochemical and petrographic studies of oldhamite, diopside, and roedderite in enstatite meteorittes. Meteoritics Planet. Sci. 33: 291–301.

Ikramuddin, M., C.M. Binz, and M.E. Lipshutz. 1976. Thermal metamorphism of primitive meteorites—II. Ten trace elements in Abee enstatite chondrite heated at 400–1000 °C. Geochim. Cosmochim. Acta, 40: 133–142.

Ikramuddin, M., M.E. Lipschutz, and E.K. Gibson. 1979. On mobile element transport in heated Abee. Meteoritics, 14: 69–80.

Imamura, K., and M. Honda. 1976. Distribution of tungsten and molybdenum between metal, silicate, and sulphide phases of meteorites. Geochim. Cosmochim. Acta, 40: 1073–1080.

Isaac, J.H., and R.R. Stewart. 1993. 3-D seismic expression of a cryptoexplosion structure. Can. J. Expl. Geophys. 29: 429–439.

Jeffery, P.M., and J.H. Reynolds. 1961. Origin of excess Xe^{129} in stone meteorites. J. Geophys. Res. 66: 3582–3583.

Jull, A.J.T., D. Lal, G.S. Burr, P.A. Bland, A.W.R. Bevan, and J.W. Beck. 2000. Radiocarbon beyond this world. Radiocarbon, 42: 151–172.

Kallemeyn, G.W., A.E. Rubin, D. Wang, and J.T. Wasson. 1989. Ordinary chondrites: bulk compositions, classification, lithophilic-element fractionations, and composition-petrographic type relationships. Geochim. Cosmochim. Acta, 53: 2747–2767.

Kaplan, I.R., and J.R. Hulston. 1966. The isotopic abundances and content of sulfur in meteorites. Geochim. Cosmochim. Acta, 30: 479–496.

Keil, K. 1968. Mineralogical and chemical relationships among enstatite chondrites. J. Geophys. Res. 73 (22): 6945–6976.

Kelso, J.A. Report of Analysis. University of Alberta, Edmonton, AB. 19 June 1939.

Kempton, R. 1996a. Abee—more questions than answers. Meteorite, 2(4): 18–19.

Kempton, R. 1996b. Abee—more questions than answers. New England Meteoretical Services, Mendon, MA. Available from http://www.meteorlab.com/METEORLAB2001dev/abeetxt. htm

Kothari, B.K., and P.S. Goel. 1974. Total nitrogen in meteorites. Geochim. Cosmochim. Acta, 38: 1493–1507.

Kracher, A., J. Willis, and J.T. Wasson. 1980. Chemical classification of iron meteorites—IX. A new group (IIF), revision of IAB and IIICD, and data on 57 additional irons. Geochim. Cosmochim. Acta, 44: 773–787.

Krinov, E.L. 1960a. Fall of Bruderheim stony meteorite shower, Canada. Meteoritical Bull. No. 18. p 1.

Krinov, E.L. 1960b. Principles of meteoritics. Translated by Irene Vidziunas. Pergamon Press, Oxford, UK.

Krouse, H.R., and R.E. Folinsbee. 1964. The S^{32}/S^{34} ratio in troilite from the Bruderheim and Peace River meteorites. J. Geophys. Res. 69: 4192–4193.

LaPaz, L. 1946. On the ownership of recovered meteorites. Pop. Astron. 54: 93–95.

LaPaz, L. 1953. Preliminary note on the Belly River, Alberta, Canada aerolite (CN=+1130,495). Meteoritics Planet. Sci. 1: 106–108.

LaPaz, L., and J. LaPaz. 1961. Space nomads. Holiday House, New York.

Laul, J.C., R. Ganapathy, E. Anders, and J.W. Morgan. 1973. Chemical fractionations in meteorites—VI. Accretion temperatures of H-, LL-, and E-chondrites, from abundance of volatile trace elements. Geochim. Cosmochim. Acta, 37: 329–357.

Lavielle, B., S. Toe, and E. Gilabert. 1997. Noble gas measurements in the L/LL5 chondrite Knyahinya. Meteoritics Planet. Sci. 32: 97–107.

Lavrukhina, A.K., and V.D. Gorin. 1979. Radioactivity of recently fallen meteorites. Meteoritics, 14: 474.

Lavrukhina, A.K., and G.K. Ustinova. 1972. Cosmogenic radionuclides in stones and meteoritic orbits. Earth Planet. Sci. Lett. 15: 347–360.

Leitch, C.A., and J.V. Smith. 1982. Petrography, mineral chemistry and origin of type I enstatite chondrites. Geochim. Cosmochim. Acta, 46: 2083–2097.

Lerbekmo, J.F., and R.M. St. Louis. 1986. The terminal Cretaceous iridium anomaly in the Red Deer Valley, Alberta, Canada. Can. J. Earth Sci. 23 :120–124.

Love, S.G., and K. Keil. 1995. Recognizing Mercurian meteorites. Meteoritics, 30: 269–278.

Loveless, A.J. 1970. Isotopic composition of gadolinium, samarium and europium in the Abee meteorite. PhD dissertation, University of British Columbia, Vancouver, BC.

Loveless, A.J., S. Yanagita, H. Mabuchi, M. Ozima, and R.D. Russell. 1972. Isotopic ratios of Gd, Sm and Eu in "Abee" enstatite chondrite. Geochim. Cosmochim. Acta, 36: 685–698.

Lovering, J.F., W. Nichiporuk, A. Chodos, and H. Brown. 1957. The distribution of gallium, germanium, cobalt, chromium, and copper in iron and stony-iron meteorites in relation to nickel content and structure. Geochim. Cosmochim. Acta, 11: 263–278.

Lowey, M. 1993. Manitou's meteorite. Calgary Herald (6 March), p. C10.

MacGregor, J.G. 1954. Behold the shining mountains. Applied Art Products, Edmonton, AB.

Malvin, D.J., D. Wang, and J.T. Wasson. 1984. Chemical classification of iron meteorites—X. Multielement studies of 43 irons, resolution of group IIIE from IIIAB, and evaluation of Cu as a taxonomic parameter. Geochim. Cosmochim. Acta, 48: 785–804.

Manuel, O.K., and M.W. Rowe. 1964. Noble gases in the Bruderheim chondrite. Geochim. Cosmochim. Acta, 28: 1999–2003.

Marti, K. 1983. Preface: the Abee consortium. Earth Planet. Sci. Lett. 62: 116–192.

Mason, B. 1963. Olivine composition in chondrites. Geochim. Cosmochim. Acta, 27: 1011–1023.

Mason, B. 1966. The enstatite chondrites. Geochim. Cosmochim. Acta, 30: 23–39.

Mason, B. 1967. Olivine composition in chondrites—a supplement. Geochim. Cosmochim. Acta, 31: 1100–1103.

Mason, B. 1971. Handbook of elemental abundances in meteorites. Gordon & Breach, New York.

Mason, B., and H.B. Wiik. 1967. The composition of the Belly River, Bluff, Bremervorde, and Modoc meteorites. Am. Mus. Novitates No. 2280: 1–19.

Masuda, H., M. Shima, and M. Honda. 1972. Distribution of uranium and thorium among components of some chondrites. Geochem. J. 6: 37–42.

Masuda, A., N. Nakamura, and T. Tanaka. 1973. Fine structures of mutually normalized rare-earth patterns of chondrites. Geochim. Cosmochim. Acta, 37: 239–248.

Matsui, T., N. Sugiura. and N.S. Brar. 1986. Gas permeability of shocked chondrites. Meteoritics, 21: 109–116.

Mazur, M.J. 1999. The seismic characterization of meteorite impact structures. MSc thesis, University of Calgary, Calgary, AB.

Mazur, M.J., and R.R. Stewart. 1998. Interpreting the Hotchkiss structure: a possible meteorite impact crater in northwestern Alberta. University of Calgary, Departrment of Geology and Geophysics, Centre for Research in Elastic Wave Exploration Seismology, Calgary, AB. CREWES Res. Rep. No. 10. pp. 35.1-35.13. Available from http://www.crewes.org/Reports/1998/1998-35.pdf

McConville, P., S. Kelley, and G. Turner. 1985. Laser probe ^{40}Ar-^{39}Ar studies of the Peace River L6 chondrite. Meteoritics, 20: 707.

McConville, P., S. Kelley, and G. Turner. 1988. Laser probe ^{40}Ar-^{39}Ar studies of the Peace River shocked L6 chondrite. Geochim. Cosmochim. Acta, 52: 2487–2499.

McCulloch, M.T., K.J.R. Rosman, and J.R. De Laeter. 1977. The isotopic and elemental abundance of ytterbium in meteorites and terrestrial samples. Geochim. Cosmochim. Acta, 41: 1703–1707.

McDougall, J. 1902. George Millward McDougall, the pioneer, patriot and missionary. 2nd ed. William Briggs, Toronto, ON.

Melcher, C.L. 1981. Thermoluminescence of meteorites and their terrestrial ages. Geochim. Cosmochim. Acta, 45: 615–626.

Melcher, C.L., and D.W. Sears. 1979. Thermal stability of thermoluminescence in chondrites. Meteoritics, 14: 249–253.

Merrihue, C. 1963. Excess xenon 129 in chondrules from the Bruderheim meteorite. J. Geophys. Res. 68: 325–330.

Merrihue, C. 1966. Xenon and krypton in the Bruderheim meteorite. J. Geophys. Res. 71: 263–313.

Miller, F.D. 1960. The meteorite. Nickelodeon, 6(3): 11–12.

Miller, S.L. 1953. A production of amino acids under possible primitive Earth conditions. Science (Washington, DC), 117: 528–529.

Millman, P.M. 1953a. A catalogue of Canadian meteorites. J. R. Astron. Soc. Can. 47: 29–32.

Millman, P.M. 1953b. Notes on Canadian meteorites. J. R. Astron. Soc. Can. 47: 162–165.

Millman, P.M. 1959. The Meanook–Newbrook meteor observatories. J. R. Astron. Soc. Can. 53(1): 15–33.

Millman, P.M. 1964. Meteor news. J. R. Astron. Soc. Can. 58: 79–82.

Millman, P.M., and D.W.R. McKinley. 1967. Stars fall over Canada. J. R. Astron. Soc. Can. 61: 277–294.

Milton, E. 1960. The Bruderheim meteorite. Stardust Magazine (Edmonton Centre. Royal Astronomical Society of Canada), April 1960.

Milton, Viscount, and W.B. Cheadle. 1865. The North-West passage by land. 4th ed. Cassell, Petter & Galpin, London.

Müller, O., P.A. Baedecker, and J.T. Wasson. 1971. Relationship between siderophilic-element content and oxidation state of ordinary chondrites. Geochim. Cosmochim. Acta, 35: 1121–1137.

Müller, O., W. Hampel, T. Kirsten, and G.F. Herzog. 1981. Cosmic-ray constancy and cosmogenic production rates in short-lived chondrites. Geochim. Cosmochim. Acta, 45: 447–460.

Murrell, M.T., and D.S. Burnett. 1982. Actinide microdistribution in the enstatite meteorites. Geochim. Cosmochim. Acta, 46: 2453–2460.

Nagai, H., T. Kobayashi, M. Honda, K. Yoshida, M. Imamura, K. Kobayashi, K. Yoshida, and H. Yamashita. 1987. Measurements of ^{10}Be and ^{26}Al in some meteorites with internal beam monitor method. Nucl. Instr. Methods Phys. Res. B29: 266–270.

Nagai, H., M. Honda, M. Imamura, and K. Kobayashi. 1993. Cosmogenic ^{10}Be and ^{26}Al in metal, carbon, and silicate of meteorites. Geochim. Cosmochim. Acta, 57: 3705–3723.

Nakamura, N., and A. Masuda. 1973. Chondrites with peculiar rare-earth patterns. Earth Planet. Sci. Lett. 19: 429–437.

Niederer, F.R., D.A. Papanastassiou, and G.J. Wasserburg. 1985. Absolute isotopic abundances of Ti in meteorites. Geochim. Cosmochim. Acta, 49: 835–851.

Nininger, A.D. 1940. Third catalog of meteorite falls (S.R.M. nos. 183–321) reported to the Society for Research on Meteorites: January 1939 to October 1940. Pop. Astron. 48: 555–560.

Nishimura, M., and E.B. Sandell. 1964. Zinc in meteorites. Geochim. Cosmochim. Acta, 28: 1055–1079.

Okazaki, R., N. Takaoka, T. Nakamura, and K. Nagao. 2000. Cosmic-ray exposure ages of enstatite chondrites. Antarct. Meteorite Res. 13: 153–169.

Oliver, T.A, 1964. Discovery of Vulcan stony meteorite, Alberta, Canada. Meteoritical Bull. No. 32.

Olsen, E., J.S. Huebner, J.A.V. Douglas, and A.G. Plant. 1973. Meteoritic amphiboles. Am. Mineral. 58: 869–872.

Olsen, E.J., and T.E. Bunch. 1984. Equilibration temperatures of the ordinary chondrites: a new evaluation. Geochim. Cosmochim. Acta, 48: 1363–1365.

Ott, U., H.P. Löhr, and J.W. Arden. 1992. Diamonds in Abee: the noble gas signature. Meteoritics, 27: 272.

Patzer, A., and L. Schultz. 2001. Noble gases in enstatite chondrites I: exposure ages, pairing, and weathering effects. Meteoritics Planet. Sci. 36: 947–961.

Patzer, A., and L. Schultz. 2002. Noble gases in enstatite chondrites II: the trapped component. Meteoritics Planet. Sci. 37: 601–612.

Pernicka, E., W. Kiesl, and M. Wichtl. 1979. The Edmonton iron. Meteoritics, 14: 515–516.

Price, G.D. 1983. The nature and significance of stacking faults in wadsleyite, natural β-(Mg, Fe)$_2$SiO$_4$ from the Peace River meteorite. Phys. Earth Planet. Interiors, 33: 137–147.

Price, G.D., A. Putnis, and D.G.W. Smith. 1982. A spinel to β-phase transformation mechanism in (Mg, Fe)$_2$SiO$_4$. Nature (London), 296: 729–731.

Price, G.D., A. Putnis, S.O. Agrell, and D.G.W. Smith. 1983. Wadsleyite, natural (Mg, Fe)$_2$SiO$_4$ from the Peace River meteorite. Can. Mineral. 21: 29–35.

Rambaldi, E.R., and M. Cendales. 1980. Siderophile element fractionation in enstatite chondrites. Earth Planet. Sci. Lett. 48: 325–334.

Rancitelli, L.A., and J.C. Laul. 1977. Cosmogenic radionuclide and trace element characterization of the Innisfree and Louisville meteorites. Meteoritics, 12: 346–347.

Reed, G.W. 1964. Fluorine in stone meteorites. Geochim. Cosmochim. Acta, 28: 1729–1743.

Reed, G.W., and R.O. Allen. 1966. Halogens in chondrites. Geochim. Cosmochim. Acta, 30: 779–800.

Reed, G.W., K. Kigoshi, and A. Turkevich. 1960. Determinations of concentrations of heavy elements in meteorites by activation analysis. Geochim. Cosmochim. Acta, 20: 122–140.

Reisener, Z., and J.I. Goldstein. 2003. Ordinary chondrite metallography: Part 2. Formation of zoned and unzoned metal particles in relatively unshocked H, L, and LL chondrites. Meteoritics Planet. Sci. 38: 1679–1696.

Reuter, J.H., S. Epstein, and H.P. Taylor. 1965. O^{18}/O^{16} ratios of some chondritic meteorites and terrestrial ultramafic rocks. Geochim. Cosmochim. Acta, 29: 481–488.

Ronaghan, A. 1973. The Iron Creek meteorite. Alta. Hist. Rev. 21(3): 10–12.

Rose, G. 1863. Systematisches verzeichnis der meteoriten den Mineralogischen Museum der Universtat zu Berlin. Ann. Phys. 118: 420–421.

Rowe, M.W., and M.A. Van Dilla. 1961. On the radioactivity of the Bruderheim chondrite. J. Geophys. Res. 66: 3553–3556.

Rowe, M.W., M.A. Van Dilla, and E.C. Anderson. 1963. On the radioactivity of stone meteorites. Geochim. Cosmochim. Acta, 27: 983–1001.

Rowe, M.W., D.D. Bogard, and O.K. Manuel. 1965. Noble gases from the Peace River, LaLande and Achilles meteorites. Geochim. Cosmochim. Acta, 29: 1199–1202.

Royal Astronomical Society of Canada. 1939. Meetings of the Society. J. R. Astron. Soc. Can. 33: 266–272.

Rubin, A.E. 1990. Kamacite and olivine in ordinary chondrites: intergroup and intragroup relationships. Geochim. Cosmochim. Acta, 54: 1217–1232.

Rubin, A.E. 1994. Metallic copper in ordinary chondrites. Meteoritics, 29: 93–98.

Rubin, A.E. 2003. Chromite-plagioclase assemblages as a new shock indicator: implications for the shock and thermal histories of ordinary chondrites. Geochim. Cosmochim. Acta, 67: 2695–2709.

Rubin, A.E. 2004. Postshock annealing and postannealing shock in equilibrated ordinary chondrites: implications for the thermal and shock histories of chondritic asteroids. Geochim. Cosmochim. Acta, 68: 673–689.

Rubin, A.E., and K. Keil. 1983. Mineralogy and petrology of the Abee enstatite chondrite breccia and its dark inclusions. Earth Planet. Sci. Lett. 62: 118–131.

Rubin, A.E., and E.R.D. Scott. 1997. Abee and related EH chondrite impact-melt breccias. Geochim. Cosmochim. Acta, 61: 425–435.

Rubin, A.E., E.R.D. Scott, and K. Keil. 1997. Shock metamorphism of enstatite chondrites. Geochim. Cosmochim. Acta, 61: 847–858.

Rushbrook, P.R., and W.D. Ehmann. 1962. Iridium in stone meteorites by neutron activation analysis. Geochim. Cosmochim. Acta, 26: 649–657.

Russell, S.S., C.T. Pillinger, J.W. Arden, M.R. Lee, and U. Ott. 1992. A new type of meteoritic diamond in the enstatite chondrite Abee. Science (Washington, DC), 256: 206–209.

Russell, W.A., D.A. Papanastassiou, and T.A. Tombrello. 1978. Ca isotope fractionation on the Earth and other solar system materials. Geochim. Cosmochim. Acta, 42: 1075–1090.

Salvatori, R., A. Maras, and E.A. King. 1984. Inventory of the Vatican meteorite collection. Meteoritics, 19: 161.

Sawatzky, H.B. 1976. Two probable late Cretaceous astroblemes in Western Canada—Eagle Butte, Alberta, and Dumas, Saskatchewan. Geophysics, 4: 1261–1271.

Schmitt, D.G. 2002. The law of ownership and control of meteorites. Meteoritics Planet. Sci. 37 (Suppl.): B5–B11.

Schmitt, R.A., R.H. Smith, J.E. Lasch, A.W. Mosen, D.A. Olehy, and J. Vasilevskis. 1963a. Abundances of the fourteen rare-earth elements, scandium, and yttrium in meteoritic and terrestrial matter. Geochim. Cosmochim. Acta, 27: 577–622.

Schmitt, R.A., R.H. Smith, and D.A. Olehy. 1963b. Cadmium abundances in meteoritic and terrestrial matter. Geochim. Cosmochim. Acta, 27: 1077–1088.

Schmitt, R.A., G.G. Goles, R.H. Smith, and T.W. Osborn. 1972. Elemental abundances in stone meteorites. Meteoritics, 7 (2): 131–213.

Schultz, L., and H. Kruse. 1989. Helium, neon, and argon in meteorites—a data compilation. Meteoritics, 24: 155–172.

Scott, E.R.D., and J.T. Wasson. 1975. Classification and properties of iron meteorites. Rev. Geophys. Space Phys. 13: 527–546.

Scott, E.R.D., J.T. Wasson, and V.F. Buchwald. 1973. The chemical classification of iron meteorites—VII. A reinvestigation of irons with Ge concentrations between 25 and 80 ppm. Geochim. Cosmochim. Acta, 37: 1957–1976.

Scott, E.R.D., J.T. Wasson, and R.W. Bild. 1977. Four new iron meteorite finds. Meteoritics, 12: 425–435.

Scott, E.R.D., J.T. Wasson, and R.W. Bild. 1978. Discovery of the Kinsella, Canada, iron meteorite. Meteoritical Bull. No. 55. Meteoritics, 13: 339–340.

Sears, D.W.G., and K.S. Weeks. 1986. Chemical and physical studies of type 3 chondrites—VI: siderophile elements in ordinary chondrites. Geochim. Cosmochim. Acta, 50: 2815–2832.

Sears, D.W.G., G.W. Kallemeyn, and J.T. Wasson. 1983. Composition and origin of clasts and inclusions in the Abee enstatite chondrite breccia. Earth Planet. Sci. Lett. 62: 180–192.

Sen Gupta, J.G. 1968. Determination of microgram amounts of the six platinum-group metals in iron and stony meteorites. Anal. Chim. Acta, 42: 481–488.

Setser, J.L., and W.D. Ehmann. 1964. Zirconium and hafnium abundances in meteorites, tektites and terrestrial materials. Geochim. Cosmochim. Acta, 28: 769–782.

Shaw, D.M. 1974. R-mode factor analysis on enstatite chondrite analyses. Geochim. Cosmochim. Acta, 38: 1607–1613.

Shedlovsky, J.P., P.J. Cressy, and T.P. Kohman. 1967. Cosmogenic radioactivities in the Peace River and Harleton chondrites. J. Geophys. Res. 72: 5051–5058.

Shima, M. 1964. The distribution of germanium and tin in meteorites. Geochim. Cosmochim. Acta, 28: 517–532.

Shima, M., and M. Honda. 1967a. Determination of rubidium-strontium age of chondrites using their separated components. Earth Planet. Sci. Lett. 2: 337–343.

Shima, M., and M. Honda. 1967b. Distributions of alkali, alkaline earth and rare earth elements in component minerals of chondrites. Geochim. Cosmochim. Acta, 31: 1995–2006.

Shima, M., and H.G. Thode. 1961. The sulfur isotope abundances in Abee and Bruderheim meteorites. J. Geophys. Res. 66: 3580.

Shimizu, M., H. Yoshida, and J.A. Mandarino. 2002. The new mineral species keilite, (Fe,Mg)S, the iron-dominant analogue of niningerite. Can. Mineral. 40: 1687–1692.

Sill, C.W., and C.P. Willis. 1962. The beryllium content of some meteorites. Geochim. Cosmochim. Acta, 26: 1209–1214.

Smith, C.L., and J.R. De Laeter. 1986. The isotopic composiiton of tellurium in the Abee meteorite. Meteoritics, 21: 133–139.

Smith, C.L., J.R. De Laeter, and K.J.R. Rosman. 1977. Mass spectrometric isotope dilution analysis of tellurium in meteorites and standard rocks. Geochim. Cosmochim. Acta, 41: 676–681.

Smith, D G.W. 1980. The mineral chemistry of the Innisfree meteorite. Can. Mineral. 18: 433–442.

Smith, D.G.W. 1997. The Ferintosh, Alberta, L6 chondrite. J. R. Astron. Soc. Can. 91: 121–126.

Smith, D.G.W., and S. Launspach. 1991. The composition of metal phases in Bruderheim (L6) and implications for the thermal histories of ordinary chondrites—erratum. Earth Planet. Sci. Lett. 102: 79–93.

Smith, D G.W., R.E. Folinsbee, and M. Hall-Beyer. 1973. A note on the mineralogy and classification of the Vilna meteorite. Meteoritics, 18: 197–199.

Southesk, Earl of. 1875. Saskatchewan and the Rocky Mountains. James Campbell & Son, Toronto, ON.

Spratt, C.E. 1989. Canada's Iron Creek meteorite. J. R. Astron. Soc. Can. 83(2): 81–91.

Srinivasan, B., R.S. Lewis, and E. Anders. 1978. Noble gases in the Allende and Abee meteorites and a gas-rich mineral fraction: investigation by stepwise heating. Geochim. Cosmochim. Acta, 42: 183–198.

Stardust Magazine. 1955a. Meteors. Royal Astronomical Society of Canada, Edmonton Centre, Edmonton, AB. Stardust Mag. 1(9&10): 1.

Stardust Magazine. 1955b. Meteors and meteorites. Royal Astronomical Society of Canada, Edmonton Centre, Edmonton, AB. Stardust Mag. 1(11&12): 1.

Steiner, J. 1981. Correction. Discovery of the Skiff, Canada, stony meteorite. Meteoritical Bull. No. 59. Meteoritics, 16: 199.

Steiner, J., and D. Smith. 1980. Discovery of the Skiff, Canada, stony meteorite. Meteoritical Bull. No. 58. Meteoritics, 15: 239.

Stöffler, D., K. Keil, and E.R.D. Scott. 1991. Shock metamorphism of ordinary chondrites. Geochim. Cosmochim. Acta, 55: 3845–3867.

Suess, H.E., and H.C. Urey. 1956. Abundances of the elements. Rev. Mod. Phys. 28: 53–74.

Suess, H.E., and H. Wänke. 1962. Radiocarbon content and terrestrial age of twelve stony meteorites and one iron meteorite. Geochim. Cosmochim. Acta, 26: 475–480.

Sugiura, N., and D.W. Strangway. 1981. The magnetic properties of the Abee meteorite: evidence for a strong magnetic field in the early system. Proc. Lunar Planet. Sci. 12B: 1243–1256.

Sugiura, N., and D.W. Strangway. 1983. A paleomagnetic conglomerate test using the Abee E4 meteorite. Earth Planet. Sci .Lett. 62: 169–179.

Tandon, S.N., and J.T. Wasson. 1968. Gallium, germanium, indium and iridium variations in a suite of L-group chondrites. Geochim. Cosmochim. Acta, 32: 1087–1109.

Tatum, J.B. 1990. A hitherto unrecorded fragment of the Bruderheim meteorite. J. R. Astron. Soc. Can. J. 84: 1–2.

Taylor, H.P., M.B. Duke, L.T. Silver, and S. Epstein. 1965. Oxygen isotope studies of minerals in stony meteorites. Geochim. Cosmochim. Acta, 29: 489–512.

Taylor, H.W. 1964a. Gamma radiation emitted by the Peace River chondrite. J. Geophys. Res. 69: 4194–4196.

Taylor, H.W. 1964b. A study of the radioactivity of the Peace River meteorite. Roy Astron. Soc. Can. J. 58: 153–162.

Thiemens, M.H., and R.N. Clayton. 1983. Nitrogen contents and isotopic ratios of clasts from the enstatite chondrite Abee. Earth Planet. Sci. Lett. 62: 165–168.

Turner, G., J.A. Miller, and R.L. Grasty. 1966. The thermal history of the Bruderheim meteorite. Earth Planet. Sci. Lett. 1: 155–157.

Turner, G., J.M. Saxton, and M. Laurenzi. 1990. Retention of K-Ar ages by meteorite fusion crust and an attempt to date Antarctic dust. Meteoritics, 25: 416.

University of Calgary Gazette. 2001. Millarville Meteorite Causes Stir in '77. University of Calgary, Calgary, AB. Univ. Calgary Gaz. 30 (22, February 20), p. 1.

Urey, H., and H. Craig. 1953. The composition of the stone meteorites and the origin of meteorites. Geochim. Cosmochim. Acta, 4: 36–82.

Urey, H., and T. Mayeda. 1959. The metallic particles of some chondrites. Geochim. Cosmochim. Acta, 17: 113–124.

Ustinova, G.K., V.A. Alekseev, and A.K. Lavrukhina. 1988. Methods for determining the preatmospheric dimensions of meteorites. [In Russian.] Geokhimiya, 1988: 1379–1395.

Van Schmus, W.R., and P.H. Ribbe. 1968. The composition and structural state of feldspar from chondritic meteorites. Geochim. Cosmochim. Acta, 32: 1327–1342.

Vogt, J.R., and W.D. Ehmann. 1965. Silicon abundances in stony meteorites by fast neutron activation analysis. Geochim. Cosmochim. Acta, 29: 373–383.

von Gunten, H.R., A. Wyttenbach, and W. Scherle. 1965. Determination of chlorine in stony meteorites by neutron activation analysis. Geochim. Cosmochim. Acta, 29: 475–480.

von Michaelis, H., L.H. Ahrens, and J.P. Willis. 1969. The composition of stony meteorites II. The analytical data and an assessment of their quality. Earth Planet. Sci. Lett. 5: 387–394.

von Michaelis, H., J.P. Willis, A.J. Erlank, and L.H. Ahrens. 1969. The composition of stony meteorites I. Analytical techniques. Earth Planet. Sci. Lett. 5: 383–386.

Voshage, H. 1978. Investigations on cosmic-ray-produced nuclides in iron meteorites, 2. New results on $^{41}K/^{40}K$-$^{4}He/^{21}Ne$ exposure ages and the interpretation of age distributions. Earth Planet. Sci. Lett. 40: 83–90.

Voshage, H. 1981. The isotopic composition of cosmic ray produced lithium in iron meteorites. Meteoritics, 16: 395.

Wacker, J.F., and K. Marti. 1983. Noble gas components in clasts and separates of the Abee meteorite. Earth Planet. Sci. Lett. 62: 147–158.

Wasson, J.T. 1969. The chemical composition of iron meteorites—III. Hexahedrites and other irons with germanium concentrations between 80 and 200 ppm. Geochim. Cosmochim. Acta, 33: 859–876.

Wasson, J.T. 1970. The chemical classification of iron meteorites IV. Irons with Ge concentrations greater than 190 ppm and other meteorites associated with group 1. Icarus, 12: 407–423.

Wasson, J.T. 1974. Meteorites: classification and properties. Springer-Verlag, New York.

Wasson, J.T., and G.W. Kallemeyn. 2002. The IAB iron-meteorite complex: a group, five subgroups, numerous grouplets, closely related, mainly formed by crystal segregation in rapidly cooling melts. Geochim. Cosmochim. Acta, 66: 2445–2473.

Wasson, J.T., and J.W. Richardson. 2001. Fractionation trends among IVA iron meteorites: contrasts with IIIAB trends. Geochim. Cosmochim. Acta, 65: 951–970.

Weber, H.W., O. Braun, L. Schultz, and F. Begemann. 1983. The noble gas record in Antarctic and other meteorites. Z. Naturforsch. 38(a): 267–272.

Weinke, H.H., R. Gijbels, R. Saelens, and W. Kiesl. 1979. Investigation of iron meteorites by ion microprobe mass analysis. Meteoritics, 14: 559–561.

Westbroek, H.-H. 1997. Seismic interpretation of two possible meteorite impact craters: White Valley, Saskatchewan and Purple Springs, Alberta. MSc thesis, University of Calgary, Calgary, AB.

Wettlaufer, B., and A. Whyte. 2004. Discovery of the Belly River meteorite: new information. J. R. Astron. Soc. Can. 98(6): 238–239.

Whipple, F.L. 1957. Meteors. Smithsonian Institution, Washington, DC. Smithsonian Rep. 1957. p. 239.

Wilson, G.C. 2004. Notes on the mineralogy of some ordinary chondrites from Alberta and Manitoba, Canada. Turnstone Geological Services Ltd., Campbellford, ON. Rep. 2004–06. pp. i–iv, 1–8.

Wilson, G.C. 2006. A compact catalogue of Canadian meteorites. Turnstone Geological Services Ltd., Campbellford, ON. Available from http://www.turnstone.ca/canamet3.pdf

Woolum, D.S., L. Bies-Horn, D.S. Burnett, and L.S. August. 1979. Bismuth and ^{208}Pb microdistributions in enstatite chondrites. Geochim. Cosmochim. Acta, 43: 1819–1828.

Wyttenbach, A., H.R. von Gunten, and W. Scherle. 1965. Determination of bromine content and isotopic composition of bromine in stony meteorites by neutron activation. Geochim. Cosmochim. Acta, 29: 467–474.

Yang, J., and S. Epstein. 1983. Interstellar organic matter in meteorites. Geochim. Cosmochim. Acta, 47: 2199–2216.

Zhai, M., and D.M. Shaw. 1994. Boron cosmochemistry. Part I: Boron in meteorites. Meteoritics, 29: 607–615.

Zhai, M., E. Nakamura, D.M. Shaw, and T. Nakano. 1996. Boron isotope ratios in meteorites and lunar rocks. Geochim. Cosmochim. Acta, 60: 4877–4881.

Zhang, Y., S. Huang, D. Schneider, P.H. Benoit, J.M. DeGart, G.F. Lofgren, and D.W.G. Sears. 1996. Pyroxene structures, cathodoluminescence and the thermal history of the enstatite chondrites. Meteoritics Planet. Sci. 31: 87–96.

Image Credits

Fig. 1.1. ©Department of Earth and Atmospheric Sciences, University of Alberta. Photograph courtesy of Andrew Locock. Used with permission.

Fig. 1.2. ©Department of Earth and Atmospheric Sciences, University of Alberta. Photograph courtesy of Andrew Locock. Used with permission.

Fig. 1.3. Drawing by the author.

Fig. 2.1. Reprinted with permission of the United Church of Canada/Victoria University Archives, Toronto, Ontario. Acc. No. 91.161P/1046 Meteorite.

Fig. 2.2. Photograph by the author.

Fig. 2.3. Photograph by the author.

Fig. 2.4. Photograph by the author.

Fig. 2.5. From Farrington, O.C. 1907. Meteorite studies II. Field Mus. Nat. Hist. Geol. Ser. Vol. 3, no. 6, plate 35. Reprinted with permission of the editor.

Fig. 2.6. From Buchwald, V.F. 1975. Handbook of iron meteorites: their history, distribution, composition, and structure. Vol. X. University of California Press, Berkeley. p. 686, fig. 930. Reprinted with permission of University of California Press.

Fig. 3.1. From Scott, E.R.D., J.T. Wasson, and R.W. Bild. 1977. Four new iron meteorite finds. Meteoritics, 12: 432. (Fig. 4). Used with permission of the editor and J.T. Wasson.

Fig. 4.1. ©Department of Earth and Atmospheric Sciences, University of Alberta. Photograph courtesy of Andrew Locock. Used with permission.

Fig. 4.2. ©Department of Earth and Atmospheric Sciences, University of Alberta. Photograph courtesy of Andrew Locock. Used with permission.

Fig. 4.3. ©Department of Earth and Atmospheric Sciences, University of Alberta. Photograph courtesy of Andrew Locock. Used with permission.

Fig. 5.1. Photograph courtesy of Laura Hayward. Used with permission.

Fig. 5.2. ©Department of Earth and Atmospheric Sciences, University of Alberta. Photograph courtesy of Andrew Locock. Used with permission.

Fig. 5.3. ©Department of Earth and Atmospheric Sciences, University of Alberta. Photograph courtesy of Andrew Locock. Used with permission.

Fig. 6.1. Photograph by Dorothy Wettlaufer. Used with permission.

Fig. 6.2. From LaPaz, L., and J. LaPaz, 1961. Space nomads. Holiday House, p. 131. Reprinted with permission of Holiday House.

Fig. 6.3. ©Department of Earth and Atmospheric Sciences, University of Alberta. Photograph courtesy of Andrew Locock. Used with permission.

Fig. 6.4. ©Department of Earth and Atmospheric Sciences, University of Alberta. Photograph courtesy of Chris Herd, Pat Cavell, and Jessica Norris; microscope and camera setup by Tom Chacko. Used with permission.

Fig. 6.5. ©Department of Earth and Atmospheric Sciences, University of Alberta. Photograph courtesy of Chris Herd, Pat Cavell, and Jessica Norris; microscope and camera setup by Tom Chacko. Used with permission.

Fig. 6.6. ©Department of Earth and Atmospheric Sciences, University of Alberta. Photograph courtesy of Chris Herd, Pat Cavell, and Jessica Norris; microscope and camera setup by Tom Chacko. Used with permission.

Fig. 7.1. Photograph by Graham C. Wilson. Used with permission.

Fig. 8.1. From Folinsbee, R.E., and L.A. Bayrock. 1961. The Bruderheim meteorite—fall and recovery. J. R. Astron. Soc. Can. 55: 219 (Fig. 2). Reprinted with permission of R.E. Folinsbee and the editor. (After Whipple 1957). Redrawn by Marvin Harder.

Fig. 8.2. From Folinsbee, R.E., and L.A. Bayrock. 1961. The Bruderheim meteorite—fall and recovery. J. R. Astron. Soc. Can. 55: 219 (Fig. 1). Reprinted with permission of R.E. Folinsbee and the editor. Redrawn by Marvin Harder.

Fig. 8.3. Provenance undetermined, from the archival record of the University of Alberta Meteorite Collection. Used with permission.

Fig. 8.4. Photograph by Stan Walker and Ty Balacko. From Folinsbee, R.E., and L.A. Bayrock. 1961. The Bruderheim meteorite—fall and recovery. J. R. Astron. Soc. Can. 55: 223 (Fig. 4). Used with permission.

Fig. 8.5. Photograph by Stan Walker and Ty Balacko. From Folinsbee, R.E., and L.A. Bayrock. 1961. The Bruderheim meteorite—fall and recovery. J. R. Astron. Soc. Can. 55: 223 (Fig. 5). Used with permission.

Fig. 8.6. From Folinsbee, R.E., and L.A. Bayrock. 1961. The Bruderheim meteorite—fall and recovery. J. R. Astron. Soc. Can. 55: 221 (Fig. 3). Reprinted with permission of R.E. Folinsbee and the editor. Redrawn by Marvin Harder.

Fig. 8.7. ©Department of Earth and Atmospheric Sciences, University of Alberta. Photograph courtesy of Chris Herd, Pat Cavell and Jessica Norris; microscope and camera setup by Tom Chacko. Used with permission.

Fig. 8.8. From Gradie, J., J. Veverka, and B. Buratti. 1980. The effects of scattering geometry on the spectrophotometric properties of powdered material. Proc. Lunar Planet, Sci. Conf. 11: 800 (Fig. 1). Used with permission.

Fig. 9.1. Photograph courtesy of the *Camrose Canadian*. Used with permission.

Fig. 9.2. ©Department of Earth and Atmospheric Sciences, University of Alberta. Photograph courtesy of Andrew Locock. Used with permission.

Fig. 9.3. ©Department of Earth and Atmospheric Sciences, University of Alberta. Photograph courtesy of Andrew Locock. Used with permission.

Fig. 9.4. From Smith, D.G.W. 1997. The Ferintosh, Alberta, L6 chondrite. J. R. Astron. Soc. Can. 91: 122 (Fig. 3). Reprinted with permission of the editor.

Fig. 10.1. From Halliday I., A.T. Blackwell, and A.A. Griffin. 1978. The Innisfree meteorite and the Canadian Camera Network. J. R. Astron. Soc. Can. 72: 29 (Fig. 4). Reprinted with permission of the editor.

Fig. 10.2. From Halliday I., A.T. Blackwell, and A.A. Griffin. 1978. The Innisfree meteorite and the Canadian Camera Network. J. R. Astron. Soc. Can. 72: 33 (Fig. 6). Reprinted with permission of the editor.

Fig. 10.3. From Halliday I., A.A. Griffin, and A.T. Blackwell. 1978. The Innisfree meteorite fall: a photographic analysis of fragmentation, dynamics and luminosity. Meteoritics, 16: 156 (Fig. 1). Reprinted with permission of the editor.

Fig. 10.4. From Halliday I., A.T. Blackwell, and A.A. Griffin. 1978, The Innisfree meteorite and the Canadian Camera Network. J. R. Astron. Soc. Can. 72: 38 (Fig. 8). Reprinted with permission of I. Halliday and the editor. Redrawn by Marvin Harder.

Fig. 10.5. ©Department of Earth and Atmospheric Sciences, University of Alberta. Photograph courtesy of Andrew Locock. Used with permission.

Fig. 10.6. ©Department of Earth and Atmospheric Sciences, University of Alberta. Photograph courtesy of Andrew Locock. Used with permission.

Fig. 11.1. From Folinsbee, R.E., and L.A. Bayrock. 1964. The Peace River meteorite: fall and recovery. J. R. Astron. Soc. Can. 58: 115 (Fig. 3). Reprinted with permission of R.E. Folinsbee and the editor. Redrawn by Marvin Harder.

Fig. 11.2. From Folinsbee, R.E., and L.A. Bayrock. 1964. The Peace River meteorite: fall and recovery. J. R. Astron. Soc. Can. 58: 119 (Fig. 5). Reprinted with permission of R.E. Folinsbee and the editor.

Fig. 11.3. From Folinsbee, R.E., and L.A. Bayrock. 1964. The Peace River meteorite: fall and recovery. J. R. Astron. Soc. Can. 58: 120 (Fig. 6). Reprinted with permission of R.E. Folinsbee and the editor. Redrawn by Marvin Harder.

Fig. 11.4. From Folinsbee, R.E., and L.A. Bayrock. 1964. The Peace River meteorite: fall and recovery. J. R. Astron. Soc. Can. 58: 122 (Fig. 9). Reprinted with permission of R.E. Folinsbee and the editor.

Fig. 11.5. ©Department of Earth and Atmospheric Sciences, University of Alberta. Photograph courtesy of Anatoly Melnik. Used with permission.

Fig. 11.6. ©Department of Earth and Atmospheric Sciences, University of Alberta. Photograph courtesy of Chris Herd, Pat Cavell and Jessica Norris; microscope and camera setup by Tom Chacko. Used with permission.

Fig. 11.7. ©Department of Earth and Atmospheric Sciences, University of Alberta. Photograph courtesy of Andrew Locock. Used with permission.

Fig. 11.8. From Chen, M., and A. El Goresy. 2000. The nature of maskelynite in shocked meteorites: not diaplectic glass but a glass quenched from shock-induced dense melt at high pressures. Earth Planet. Sci. Lett. 179: 491 (Fig. 1b). Reprinted with permission from Elsevier.

Fig. 11.9. From Chen, M., and A. El Goresy. 2000. The nature of maskelynite in shocked meteorites: not diaplectic glass but a glass quenched from shock-induced dense melt at high pressures. Earth Planet. Sci. Lett. 179: 493 (Fig. 3b). Reprinted with permission from Elsevier.

Fig. 11.10. From Chen, M., and A. El Goresy. 2000. The nature of maskelynite in shocked meteorites: not diaplectic glass but a glass quenched from shock-induced dense melt at high pressures. Earth Planet. Sci. Lett. 179: 494 (Fig. 4a). Reprinted with permission from Elsevier.

Fig. 12.1. ©Department of Earth and Atmospheric Sciences, University of Alberta. Photograph courtesy of Andrew Locock. Used with permission.

Fig. 12.2. ©Department of Earth and Atmospheric Sciences, University of Alberta. Photograph courtesy of Andrew Locock. Used with permission.

Fig. 12.3. ©Department of Earth and Atmospheric Sciences, University of Alberta. Photograph courtesy of Chris Herd, Pat Cavell and Jessica Norris; microscope and camera setup by Tom Chacko. Used with permission.

Fig. 12.4. ©Department of Earth and Atmospheric Sciences, University of Alberta. Photograph courtesy of Andrew Locock. Used with permission.

Fig. 13.1. From Folinsbee, R.E., L.A. Bayrock, G.L. Cumming, and D.G.W. Smith. 1969. Vilna meteorite—camera, visual, seismic and analytic records. J. R. Astron. Soc. Can. 63: 62 (Fig. 1). Reprinted with permission of R.E. Folinsbee and the editor. Redrawn by Marvin Harder.

Fig. 13.2. From Folinsbee, R.E., L.A. Bayrock, G.L. Cumming, and D.G.W. Smith. 1969. Vilna meteorite—camera, visual, seismic and analytic records. J. R. Astron. Soc. Can. 63: 63 (Fig. 2). Reprinted with permission of R.E. Folinsbee and the editor.

Fig. 13.3. From Folinsbee, R.E., L.A. Bayrock, G.L. Cumming, and D.G.W. Smith. 1969. Vilna meteorite—camera, visual, seismic and analytic records. J. R. Astron. Soc. Can. 63: 65 (Fig. 3). Reprinted with permission of R.E. Folinsbee and the editor. Redrawn by Marvin Harder.

Fig. 13.4. From Folinsbee, R.E., L.A. Bayrock, G.L. Cumming, and D.G.W. Smith. 1969. Vilna meteorite—camera, visual, seismic and analytic records. J. R. Astron. Soc. Can. 63: 75 (Fig. 9). Reprinted with permission of R.E. Folinsbee and the editor. Redrawn by Marvin Harder.

Fig. 13.5. From Folinsbee, R.E., L.A. Bayrock, G.L. Cumming, and D.G.W. Smith. 1969. Vilna meteorite—camera, visual, seismic and analytic records. J. R. Astron. Soc. Can. 63: 66 (Fig. 4). Reprinted with permission of R.E. Folinsbee and the editor. Redrawn by Marvin Harder.

Fig. 13.6. From Folinsbee, R.E., L.A. Bayrock, G.L. Cumming, and D.G.W. Smith. 1969. Vilna meteorite—camera, visual, seismic and analytic records. J. R. Astron. Soc. Can. 63: 80 (Fig. 12). Reprinted with permission of R.E. Folinsbee and the editor.

Fig. 13.7. From Folinsbee, R.E., L.A. Bayrock, G.L. Cumming, and D.G.W. Smith. 1969. Vilna meteorite—camera, visual, seismic and analytic records. J. R. Astron. Soc. Can. 63: 81 (Fig. 13). Reprinted with permission of R.E. Folinsbee and the editor.

Fig. 14.1. ©Department of Geology and Geophysics, University of Calgary. Photograph courtesy of Carmen Chinery and Larry R. Lines. Used with permission.

Fig. 14.2. ©Department of Earth and Atmospheric Sciences, University of Alberta. Photograph courtesy of Andrew Locock. Used with permission.

Fig. 15.1. From Griffin, A.A., P.M. Millman, and I. Halliday. 1992. The fall of the Abee meteorite and its probable orbit. J. R. Astron. Soc. Can. 86: 8 (Fig. 1). Reprinted with permission of the editor.

Fig. 15.2. Geological Survey of Canada specimen #GSC 109341B. From Dawson, K.R., J.A. Maxwell, and D.E. Parsons. 1960. A description of the meteorite that fell near Abee, Alberta, Canada. Geochim. Cosmochim. Acta, 21: following p. 127, plate 1. Reproduced with the permission of the Minister of Public Works and Government Services Canada, 2005 and courtesy of Natural Resources Canada, Geological Survey of Canada; also with permission from Elsevier.

Fig. 15.3. Geological Survey of Canada specimen #GSC 109341C. From Dawson, K.R., J.A. Maxwell, and D. E. Parsons. 1960. A description of the meteorite that fell near Abee, Alberta, Canada. Geochim. Cosmochim. Acta, 21: following p. 127, plate 2. Reproduced with the permission of the Minister of Public Works and Government Services Canada, 2005 and courtesy of Natural Resources Canada, Geological Survey of Canada; also with permission from Elsevier.

Fig. 15.4 From Marti, K. 1983. Preface: the Abee Consortium. Earth Planet. Sci. Lett. 62: 116 (Fig. 1). Reprinted with permission from Elsevier and the author.

Fig. 15.5 From Marti, K. 1983. Preface: the Abee Consortium. Earth Planet. Sci. Lett. 62: 117 (Fig. 1). Reprinted with permission from Elsevier and the author.

Fig. 15.6 ©Department of Earth and Atmospheric Sciences, University of Alberta. Photograph courtesy of Chris Herd. Used with permission.

Fig. 16.1 From Millman, P.M. 1959. The Meanook–Newbrook meteor observatories. J. R. Astron. Soc. Can. 53: 17 (Fig. 1). Reprinted with permission of the editor.

Fig. 16.2. From Millman, P.M. 1959. The Meanook–Newbrook meteor observatories. J. R. Astron. Soc. Can. 53: 23 (Fig. 7). Reprinted with permission of the editor.

Fig. 16.3. From Millman, P.M. 1959. The Meanook–Newbrook meteor observatories. J. R. Astron. Soc. Can. 53: 24 (Fig. 8). Reprinted with permission of the editor.

Fig. 16.4. From Millman, P.M. 1959. The Meanook–Newbrook meteor observatories. J. R. Astron. Soc. Can. 53: 28 (Fig. 10). Reprinted with permission of the editor.

Fig. 16.5. Photograph by the author.

Fig. 16.6. Photograph by the author.

Fig. 16.7. Photograph by the author.

Fig. 16.8. From Halliday, I., A.T. Blackwell, and A.A. Griffin. 1978. The Innisfree meteorite and the Canadian Camera Network. J. R. Astron. Soc. Can. 72: 18 (Fig. 1). Reprinted with permission of I. Halliday and the editor. Redrawn by Marvin Harder.

Fig. 16.9. From Halliday, I., A.T. Blackwell, and A.A. Griffin. 1978, The Innisfree meteorite and the Canadian Camera Network. J. R. Astron. Soc. Can. 72: 19 (Fig. 2). Reprinted with permission of I. Halliday and the editor.

Fig. 16.10. Photograph courtesy of Blaise MacMullin, Athabasca University. Used with permission.

Fig. 16.11. Photographs courtesy of Athabasca University and Alister Ling. Used with permission.

Fig. 17.1. Photograph courtesy of Franklin Loehde. Used with permission.

Fig. 17.2. ©Brad Gledhill. Used with permission.

Fig. 18.1. Photograph by the author.

Fig. 18.2. Photograph courtesy of Tanya Whyte and Jerome Martin. Used with permission.

Fig. 18.3. NAPL photograph #A27794–44. ©2006. Reproduced under license from Her Majesty the Queen in Right of Canada, with permission of Natural Resources Canada.

Fig. 18.4. Photograph courtesy of Mark Zalcik. Used with permission.

Fig. 18.5. From Mazur, M.J., R.S. Stewart, A.R. Hildebrand, D.C. Lawton, and H.-H. Westbroek. 1999. Seismic characterization of impact craters: a summary. University of Calgary, Department of Geology and Geophysics, Centre for Research in Elastic Wave Exploration Seismology, Calgary, Alberta. CREWES Res. Rep. No. 11. p. 54.12, fig. 6. Image by Henry Bland, University of Calgary. Used with permission.

Fig. 19.1. Photograph courtesy of Alan Zalaski. Used with permission.

Fig. 19.2. Photograph courtesy of Alan Zalaski. Used with permission.

Fig. 19.3. Photograph courtesy of Alan Zalaski. Used with permission.

Index

Page numbers in italics refer to illustrations or figures.

Edberg, 206
Edmonton, *xvi*, 1, 14, 15, 18, 57–63, 66, 87, 99,
102, 141–43, 145, 147, 151, 155, 156,
193, 195, *201*, 203, 205, 219
Edmonton (Canada) meteorite, xx, 1–9, 21,
24, 33, 40, 44, 230n1.10, 242n14.2
Edmonton Centre (RASC), 1, 57, 60, 62, 204,
205, 206, 246n17.2
Edmonton (Kentucky) meteorite, 7
Edmonton, Mr. and Mrs. Dan, 143
Edmonton Space Sciences Centre, 205
Egan, W.G., 70, 71
Egremont, 58, *59*, 65, *66*, 103, *117*
EH chondrites, 78, 157, 163, 166, 167, 222, 223
Ehmann, William D., 71, 73, 75–77, 82, 127,
128, 171, 173, 175, 177, 178
Eidg. Institut für Reaktorforschung, 80, 183
EL chondrites, 78, 157, 166, 222, 223
El Goresy, Ahmed, 124, 125
electron microprobe analysis, xxii, 9, 53, 54,
67, 94, 96, 106, 108, 109, 126, 149,
166, 250
electron microscopy, 124
ellipse of fall, 32, *59*, 65, *66*, 103, *117*, 120,
121, *143*, *147*
EMP. *See* electron microprobe analysis
Enarson, Donald A., 93, 94
Enarson, I. Samuel, 93. *94*
Enarson, Mrs., 93
Endiang, 206
energy-dispersive spectrometer, 97, 106
Englert, P., 86, 90, 134, 135, 188
Enrico Fermi Institute, 183
enstatite, xix, 96, 157, 158, 163, 166, 167, 249,
250
enstatite chondrites, xix, xx, xxiii, 9, 78, 134,
156–58, 162–67, 170, 176–79, 182–85,
188, 189, 209, 222, 223, 242n15.10,
249, 250
Entwistle, 145
Epstein, Samuel, 88, 184–86
Er. *See* erbium
Erasmus, Wayne, 147
erbium, 72, 74, 76, 130, 132, 174
Eu. *See* europium
European Network, 198
europium, 72, 74, 76, 111, 130, 132, 172–74
Evensen, N.M., 170
excentroradial chondrules, 51, 54, 106, 249
exposure age, xxii, 6, 8, 9, 33–36, 84, 86–91,
112, 122, 133–35, 187, 188, 222

F. *See* fluorine
Fa. *See* fayalite
facies, 70, 127, 249
Farrington, O.C., 21
fast neutrons, 132, 135, 186
fayalite, xix, 47, 53, 55, 67, 96, 137, 139, 151,
249, 250
Fe. *See* iron
Fedechko, William, 102, 103
feldspar, xvii, 50, 52, 55, 67, 68, 85, 86, 94, 96,
110, 125–27, 154, 213
Ferguson, Lyell, 100
Ferguson, Martin, 100
Ferintosh, *xvi*, 93
Ferintosh meteorite, *xvi*, xx, 53, 55, 93–97
ferrosilite, xix, 96, 229nI.6, 249
Field Columbian Museum, 21
Field Museum of Natural History, 24, 69, 167
fireball, 27, 58, 60, 62, 66, 99–101, 104, 105,
115, 116, 118, 141, 142, 144–47, 155,
156, 194, 196, 200, 201, *203*, 206, *207*,
209, 212, 213
Fireman, Edward, 83, 86, 88, 89
fission, 81, 132, 187
fission track, 81, 187
fission track radiography, 82, 178–81, 188
Flagstaff Hill, 17
Flat Lake, *143*
fluorapatite, 81
fluorine, 75, 80, 81, 96, 97, 131, 133, 173, 183,
249, 250
fluorometry, 82, 176, 180, 181
Fo. *See* forsterite
Folinsbee, R.E., 5, 7, 9, 18, 19, 31–33, 46,
62–67, 71, 78, 82, 88, 89, 91, 93, 99,
102, 112, 115, 118, 120, 121, 127,
129, 134, 136, 137, 141, 149, 207,
230n1.15, 234n8.11, 235n8.33,
236n8.78, 241n12.3, 242n14.8
Fork Lake, 206
formation age. *See* solidification age
forsterite, xix, 47, 137, 249, 250
Fort Macleod, *xvi*, 43, 44, 53, 233n6.6
Fort McMurray, *xvi*, 141
Fort Victoria, 41
Fouche, K.F., 173, 175, 176
Fox Lake, 206
fractional crystallization, 35
fractional dissolution, 130
fragmentation, 135, 200, 223
frame-grabber technology, 204

60, 67, 68, 73, 77, 78, 89, 91, 110, 121, 122, 128, 137, 139, 158, 161, 166, 168, 170, 220, 222, 229nI.2, 234n8.2
Niederer, F.R., 179
Nihon University, 79, 135
Nininger, Harvey, H. 27, 142, 162, 167, 220, 250
niningerite, 162, 166, 167, 178, 250
Nishimura, Masakichi, 75, 78, 173, 177
nitrogen, 75, 80, 83, 131, 133, 171, 183, 184, 185
noble gas, 79, 80, 87, 90, 91, 132, 134, 135, 177, 180, 182, 187, 222, 223
Noland, M., 70
non-magmatic, 35, 40
non-magnetic, 68, 86, 132, 172, 176, 177, 179, 184
Norberg Lake, *145*, 149
North Saskatchewan River, *xvi*, 15, 64, 155
Northwest Territories, 37, 245n15.113
Norwalk, Connecticut, 194
NRC. *See* National Research Council of Canada
NRM. *See* natural remanent magnetization
nuclear tracks, 187
nucleus, 49
nuclides, 34, 35, 187, 189

O. *See* oxygen
Observatoire de Paris, 167
octahedrite, xviii, xx, 7, 21, 23–25, 28, 31, 35, 209
Okayama University, 80, 110, 133, 178
oldhamite, 157, 158, 161, 162, 178, 179, 182, 186, 222, 243n15.17
Oldman River, 43–45, 55
Olehy, D.A., 178
Oliver, T.A., 151
olivine, xviii–xx, 47, 49–52, 53–55, 67, 68, 81, 83, 85, 86, 94, 96, 106, 108, 109, 112, 121, 122, 124, 125, 127, 137, 149, 151, 167, 185, 249, 250
Olsen, Edward J., 69, 70, 167
O-mah-chees. *See* Whitford, John
Onushko, Mr., 64
O'Nions, R.K., 170
Open University, 184
Operation WHAM, 205
optical microscopy, 94, 166, 230n1.14, 250
orthopyroxene, 51, 52, 54, 55, 69, 94, 96, 122, 149, 163, 229nI.6

Os. *See* osmium
Osborn, T.W., 71, 127, 168
osmium, 7, 73, 74, 77, 111, 129, 133, 175, 177, 178
Ott, U., 184, 185
Ottawa, Ontario, 24, 44, 63, 67, 101, 102, 106, 129, 137, 151, 156, 158, 165, 167, 193, 194, 197, 198, 212, 213, 235n8.33
Ouyang, Xinwei, 35
Oxford University, 85
oxygen, 71, 81–83, 113, 133, 157, 177, 184–86
Ozima, M., 173

P. *See* phosphorus
Page, Arthur, 146, *196*
Pakan, 15
paleomagnetism, 162
Palisades, New York, 170
palladium, 77, 129, 132, 174
Papanastassiou, D.A., 179
Papia, Grace M., 73
Paradis, Gustav, 144
Parsons, D.E., 158
Pasadena, California, 67, 83, 184, 185
Patzer, Andrea, 222, 223
Paul's Band Indian Reserve, 57
Pb. *See* lead
Pd. *See* palladium
Peace River (town), *xvi*, 116, 117
Peace River meteorite, xvi, xx, 55, 66, 72, 77, 80–82, 84, 90, 94, 109, 110, 115–36, 137, 149, 167, 188, 205, 250
Peabody Museum, 65
Pennsylvania State University, 5, 185
Perihelion, 57, 105
Perkin-Elmer Corporation, 194
Pernicka, Ernst, 7, 9
Peters, Bill, 17, 18
PGNAA. *See* prompt gamma neutron activation analysis
phase, xxi, 6, 9, 69, 72, 74, 77, 80–82, 84, 85, 91, 107, 109, 124, 130, 131, 136, 164, 169, 174, 177–79, 182–84, 186, 250
phenocryst, 54, 250
phosphate, 68, 72, 81, 106, 149, 249, 250
phosphide, 5, 6
phosphorus, 3, 6, 73, 171, 176
Physical Research Laboratory, 187
Peigan, 44
Pierce, 43
piezoglyphs. *See* regmaglypts

Fig. 6.4. A polished thin section of the Belly River meteorite seen in transmitted plane-polarized light. Opaque shock veins can be seen running diagonally across the field. (35× magnification). Field of view 3.3 mm.

Fig. 6.5. A polished thin section of the Belly River meteorite seen in transmitted cross-polarized light. The large green mineral is olivine; note the shock-induced fracturing (42× magnification). Field of view 2.7 mm.

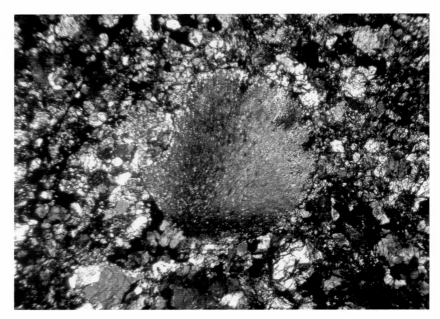

Fig. 6.6. A radial pyroxene chondrule in the Belly River meteorite (cross-polarized light). Recrystallization has partially obliterated the margin of the chondrule. (32× magnification). Field of view 3.6 mm.

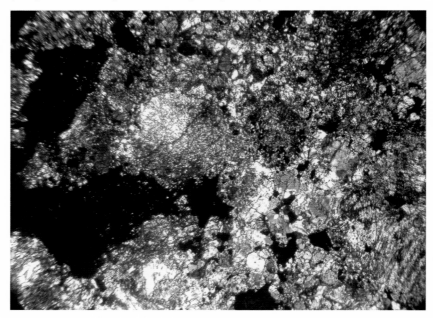

Fig. 8.7. A polished thin section of the Bruderheim meteorite seen in transmitted cross-polarized light. The large chondrule near the centre shows signs of shock; the adjacent smaller, darker chondrule is nearly intact. Note the opaque minerals (black). (31× magnification). Field of view 4.9 mm.

Fig. 11.6. The recrystallized structure of the Peace River meteorite, with shock veins, seen in transmitted cross-polarized light. (30× magnification). Field of view 4.9 mm.

Fig. 12.3. Two chondrules (centre-left) and a hexagon-shaped crystal of fayalite (centre-right) in the Skiff meteorite seen in cross-polarized light. (25× magnification). Field of view 4.5 mm.

Fig. 15.6 A polished thin section of the Abee meteorite viewed in reflected light. The light areas are highly reflective nickel-iron alloy. Note the round relict chondrule near bottom right. (40× magnification).